Coral Reefs Philippines

Reef ID Books

Photographic guide with 1600 species covered

A.S. Ryanskiy

INTRODUCTION, COPYRIGHT, ACKNOWLEDGMENTS

INTRODUCTION

Philippines belong to the coral triangle, an area with more species of fish and corals than any other marine environment on earth. Inside Coral Reefs Philippines:
• The most comprehensive photo guide to Philippine marine life, covering fishes, turtles, invertebrates and marine plants
• Designed for divers, snorkelers and marine life lovers
• Features near 1600 species, among them new species and new records of fishes and invertebrates for the Philippines
• All photographs taken in natural environment, no anaesthetic or sedatives substances were used on animals
• Small and compact enough to be taken for your next dive trip

COPYRIGHT

COPYRIGHT ©2018 Andrey Ryanskiy. First edition. All rights reserved. All images © Andrey Ryanskiy unless otherwise stated.

ACKNOWLEDGMENTS

The author wishes to thank all those who helped to identify all the 1600+ coral reef organisms, featured in this book. Among them are world known scientists, and just people in love with the marine life. All correct ID in this book - it is their merit, all the mistakes on me:

Frédéric Ducarme, Joe Rowlett, Sergey Bogorodskiy, Leslie Harris, Ondřej Radosta, Zdenek Duris, Onafets Amsiard, Arthur Anker, Jose Cristopher Mendoza, Machel D.Malay, Marli Wakeling, Nathaniel Evans, Christopher L. Mah, Jim Greenfield, David Paul Speed, Ron Silver, Anthony Gill, Hsini Lin.

A special thanks to my dive guides and friends Peri Paleracio and William Mendoza.

I would like to thank my wife Irina Khlopunova for continuously and patiently supporting me in every phase of the work on this book.

ABBREVIATIONS

ID - Identification tips

sp. - used when the actual specific name cannot or need not be specified

cf. - used to indicate undescribed species assumed related to, but distinct from a described species

IP, IWP, WP - Indo-Pacific, Indo-West Pacific, West Pacific

DBA - discovered by author, IT - Identification Tentative

CONTENTS

BONY FISHES 4
SHARKS 66
RAYS 67
SNAKES 67
TURTLES 67
CRUSTACEANS 68
Shrimps 68
Prawns 81
Rock shrimps 82
Boxer shrimps 82
Lobsters 83
Mantis shrimps 84
Sea spiders 85
Squat lobsters 86
Hermit crabs 87
Porcelain crabs 90
True crabs 91
Amphipods 102

Copepods 102
Isopods 103
Barnacles 103
MOLLUSCS 104
Sea snails 104
Chitons 114
Bivalves 114
Nudibranchs 116
Cuttlefishes 160
Squids 160
Octopuses 160
ECHINODERMS 162
Sea stars 162
Brittle stars 165
Feather stars 166
Sea urchins 167
Sea cucumbers 169

MARINE WORMS 171
Flat worms.............................171
Ribbon worms.......................176
Peanut worms.......................176
Fire worms 176
Bobbit worms 177
Tube-building worms 177
Polynoid worms 178
Tube worms 179
Spaghetti worms 180
Horseshoe worms 180
SEA SQUIRTS 181
SPONGES 185
BRYOZOANS 187
CNIDARIANS 188
Soft corals..............................188
Sea fans, sea pens.................189
Anemones...............................190
Corallimorphs 191
Zoanthids 191
Tube anemones 192
Lace corals 192
Hydroids, fire corals 193
Hard corals 193
Jellyfishes 198
COMB JELLIES 198
MARINE PLANTS.................199

BONY FISHES — MORAY EELS (MURAENIDAE)

Zebra moray (*Gymnomuraena zebra*)
Moray eels (Muraenidae)
Indo-Pacific, 150 cm

Fimbriated moray (*Gymnothorax fimbriatus*)
Moray eels (Muraenidae) Indo-Pacific, 80 cm

Yellowmargin moray (*Gymnothorax flavimarginatus*)
Moray eels (Muraenidae)
Indo-Pacific, 240 cm

Whitemouth moray (*Gymnothorax meleagris*)
Moray eels (Muraenidae) Indo-Pacific, 120 cm
ID: Inside of mouth white

White eyed moray (*Gymnothorax thyrsoideus*)
Moray eels (Muraenidae)
Indo-Pacific, 66 cm

Richardson's moray (*Gymnothorax richardsonii*)
Moray eels (Muraenidae)
Indo-Pacific, 34 cm

Bar-tail moray (*Gymnothorax zonipectus*)
Moray eels (Muraenidae)
Indo-Pacific, 50 cm

Whitemargin moray (*Gymnothorax albimarginatus*)
Moray eels (Muraenidae)
Indo-Pacific, 105 cm

BONY FISHES | MORAY EELS (MURAENIDAE)

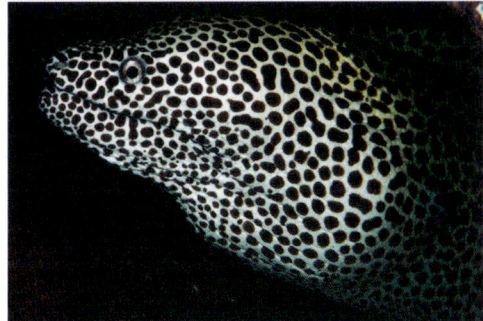

Honeycomb moray (*Gymnothorax favagineus*)
Moray eels (Muraenidae) Indo-Pacific, 300 cm

White ribbon eel (*Pseudechidna brummeri*)
Moray eels (Muraenidae) Indo-Pacific, 103 cm
ID: head with small dark spots

Blackcheek moray (*Gymnothorax breedeni*)
Moray eels (Muraenidae) Indo-Pacific, 75 cm
ID: brown with black blotches

Ribbon eel (*Rhinomuraena quaesita*)
Moray eels (Muraenidae) Indo-Pacific, 120 cm
ID: males are blue, females yellow, juveniles black

Reptilian snake eel (*Brachysomophis henshawi*)
Snake eels (Ophichthidae) Indo-Pacific, 120 cm

Tiger snake eel (*Myrichthys maculosus*)
Snake eels (Ophichthidae)
Indo-Pacific, 100 cm

Napoleon snake eel (*Ophichthus bonaparti*)
Snake eels (Ophichthidae)
Indo-Pacific, 75 cm

BONY FISHES | CONGER EELS (CONGRIDAE)

Dark-shouldered snake eel (*Ophichthus cephalozona*) Snake eels (Ophichthidae)
Pacific Ocean, 115 cm

Many-eyed snake eel (*Ophichthus polyophthalmus*)
Snake eels (Ophichthidae)
Indo-Pacific, 63 cm

Shleele's conger (*Ariosoma shleeli*)
Conger eels (Congridae) Indo-West Pacific, 25 cm
ID: row of white dots under dorsal fin

Longfin african conger (*Conger cinereus*)
Conger eels (Congridae)
Indo-Pacific, 140 cm

Barnes' garden eel (*Gorgasia barnesi*)
Conger eels (Congridae)
West Pacific, 121 cm

Splendid garden eel (*Gorgasia preclara*)
Conger eels (Congridae)
Indo-West Pacific, 40 cm

Spotted garden-eel (*Heteroconger hassi*)
Conger eels (Congridae)
Indo-Pacific, 50 cm

Taylor's garden eel (*Heteroconger taylori*)
Conger eels (Congridae)
Western Central Pacific, 48 cm

BONY FISHES — LIZARDFISHES, SOLDIERFISHES

Striped eel catfish (*Plotosus lineatus*)
Eeltail catfishes (Plotosidae)
Indo-Pacific, 32 cm

Two-spot lizard fish (*Synodus binotatus*)
Lizardfishes (Synodontidae)
Indo-Pacific, 18 cm

Lighthouse lizardfish (*Synodus jaculum*)
Lizardfishes (Synodontidae) Indo-Pacific, 20 cm
ID: black spot on caudal fin base

Tectus lizardfish (*Synodus tectus*)
Lizardfishes (Synodontidae)
Indo-West Pacific, 10 cm

Reef lizardfish (*Synodus variegatus*)
Lizardfishes (Synodontidae) Indo-West Pacific, 40 cm
ID: intermittent white stripe on lower side

Shadowfin soldierfish (*Myripristis adusta*)
Soldierfishes (Holocentridae)
Indo-Pacific, 35 cm

Whitetip soldierfish (*Myripristis vittata*)
Soldierfishes (Holocentridae)
Indo-Pacific, 25 cm

Sammara squirrelfish (*Neoniphon sammara*)
Soldierfishes (Holocentridae)
Indo-Pacific, 15 cm

BONY FISHES — SHRIMPFISHES, GHOST PIPEFISHES

Threespot squirrelfish (*Sargocentron cornutum*)
Squirelfishes (Holocentridae)
Indo-West Pacific, 27 cm

Crown squirrelfish (*Sargocentron diadema*)
Squirelfishes (Holocentridae) Indo-Pacific, 17 cm
ID: dorsal fin with white spine tips

Short dragonfish (*Eurypegasus draconis*)
Seamoths (Pegasidae)
Indo-Pacific, 10 cm

Grooved razor-fish (*Centriscus scutatus*)
Shrimpfishes (Centriscidae)
Indo-Pacific, 15 cm

Razorfish (*Aeoliscus strigatus*)
Shrimpfishes (Centriscidae)
Indo-West Pacific, 15 cm

Robust ghost pipefish (*Solenostomus cyanopterus*) Ghost pipefishes (Solenostomidae)
Red Sea, Indo-Pacific, 17 cm

BONY FISHES   PIPEFISHES & SEAHORSES (SYNGNATHIDAE)

Velvet ghost pipefish (*Solenostomus* (cf?) *cyanopterus*) Ghost pipefishes (Solenostomidae) Philippines, 10 cm, rare colour variation

Halimeda ghostpipefis (*Solenostomus halimeda*) Ghost pipefishes (Solenostomidae) Indo-Pacific, 5 cm

Harlequin ghost pipefish (*Solenostomus paradoxus*) Ghost pipefishes (Solenostomidae) Red Sea, Indo-Pacific, 12 cm

Shortpouch pygmy pipehorse (*Acentronura breviperula*) Pipefishes & seahorses (Syngnathidae) Indo-Pacific, 5 cm

Pugheaded pipefish (*Bulbonaricus davaoensis*) Pipefishes & seahorses (Syngnathidae) Indo-West Pacific, 4 cm

Ringed pipefish (*Dunckerocampus dactyliophorus*) Pipefishes & seahorses (Syngnathidae) Indo-Pacific, 19 cm. **ID**: tail with white spot in the centre

Kulbicki's pipefish (*Festucalex* cf. *kulbickii*) Pipefishes & seahorses (Syngnathidae) West Pacific, 7 cm

BONY FISHES   PIPEFISHES & SEAHORSES (SYNGNATHIDAE)

Winged seahorse (Hippocampus alatus)
Pipefishes & seahorses (Syngnathidae)
Indo-West Pacific, 12 cm

Pygmy seahorse (*Hippocampus bargibanti*)
Pipefishes & seahorses (Syngnathidae)
Indo-West Pacific, 2,4 cm

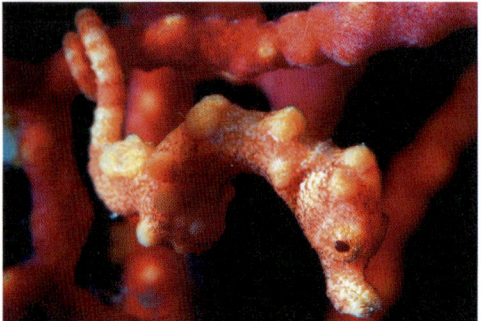

Denise's pygmy seahorse (*Hippocampus denise*)
Pipefishes & seahorses (Syngnathidae)
West Pacific, 2,2 cm

Thorny seahorse (*Hippocampus histrix*)
Pipefishes & seahorses (Syngnathidae)
Indo-Pacific, 17 cm

Great seahorse (*Hippocampus kelloggi*)
Pipefishes & seahorses (Syngnathidae)
West Pacific, 28 cm

Mollucan seahorse (*Hippocampus moluccensis*)
Pipefishes & seahorses (Syngnathidae)
Indonesia, 16 cm

Pontoh's pygmy seahorse (*Hippocampus pontohi*) Pipefishes & seahorses (Syngnathidae)
West Pacific, 1,7 cm. **ID**: Whitish with scattered red filaments. **Right picture**: Severnsi pygmy seahorse, colour variation. Usually darker and smaller, 5-8 mm

BONY FISHES — FLYING GURNADS, LIONFISHES

Trunk-barred pipefish (*Phoxocampus tetrophthalmus*)
Pipefishes & seahorses (Syngnathidae)
Western Central Pacific, 8 cm

Softcoral pipefish (*Siokunichthys breviceps*)
Pipefishes & seahorses (Syngnathidae)
Indo-West Pacific, 12 cm

Mushroom-coral pipefish (*Siokunichthys nigrolineatus*)
Pipefishes & seahorses (Syngnathidae) WCP, 8 cm
ID: dark stripe through eye

Double-ended pipefish (*Trachyrhamphus bicoarctatus*)
Pipefishes & seahorses (Syngnathidae)
Indo-West Pacific, 40 cm

Oriental flying gurnar (*Dactyloptena orientalis*)
Flying gurnards (Dactylopteridae)
Indo-Pacific, 40 cm

Twospot turkeyfish (*Dendrochirus biocellatus*)
Lionfishes (Scorpaenidae)
Indo-Pacific, 13 cm

Shortfin turkeyfish (*Dendrochirus brachypterus*)
Lionfishes (Scorpaenidae)
Indo-West Pacific, 17 cm

Zebra lionfish (*Dendrochirus zebra*)
Lionfishes (Scorpaenidae)
Indo-West Pacific, 25 cm

BONY FISHES — LIONFISHES (SCORPAENIDAE)

Mozambique scorpionfish (*Parascorpaena mossambica*) Lionfishes (Scorpaenidae) Indo-Pacific, 12 cm

Northern scorpionfish (*Parascorpaena picta*) Lionfishes (Scorpaenidae) Indo-West Pacific, 16 cm
ID: bars on lips, red bands on iris

Ruby scorpionfish (*Parascorpaena* sp.) Lionfishes (Scorpaenidae) Philippines, 20 cm
New species, discovered by author

Broadbarred firefish (*Pterois antennata*) Lionfishes (Scorpaenidae) Indo-Pacific, 20 cm
ID: dark spots near the base of pectoral fins

Red lionfish (*Pterois russellii*) Lionfishes (Scorpaenidae) Pacific Ocean, 38 cm
ID: fins with dark spots, often tentacles above eyes

Sculpin scorpionfish (*Scorpaenopsis cotticeps*) Lionfishes (Scorpaenidae) Indo-West Pacific, 6,3 cm

Paddle-flap scorpionfish (*Rhinopias eschmeyeri*) Lionfishes (Scorpaenidae) Indo-West Pacific, 23 cm.
ID: two tentacles on the underside of the lower jaw

BONY FISHES | LIONFISHES, FLATHEADS

Weedy scorpionfish (*Rhinopias frondosa*) Lionfishes (Scorpaenidae)
Indo-West Pacific, 20 cm.
ID: variable colours, 9-12 tentacles on the underside of the lower jaw

Tassled scorpionfish (*Scorpaenopsis oxycephala*)
Lionfishes (Scorpaenidae)
Indo-West Pacific, 30 cm

Eyebrow scorpionfish (*Sebastapistes taeniophrys*)
Lionfishes (Scorpaenidae)
Indo-West Pacific, 2,5 cm

Leaf scorpionfish (*Taenianotus triacanthus*)
Lionfishes (Scorpaenidae)
Indo-Pacific, 10 cm

Crocodile flathead (*Cymbacephalus beauforti*)
Flatheads (Platycephalidae)
West Pacific, 70 cm

Welander's flathead (*Rogadius welanderi*)
Flatheads (Platycephalidae) Indo-West Pacific, 15 cm
ID: fins with yellow spots and narrow white margin

Longsnout flathead (*Thysanophrys chiltonae*)
Flatheads (Platycephalidae)
Indo-Pacific, 25 cm

BONY FISHES   VELVETFISHES, WASPFISHES, STONEFISHES

Wasp-spine velvetfish (*Acanthosphex leurynnis*)
Velvetfishes (Aploactinidae)
Indo-West Pacific, 3 cm

Dusky velvetfish (*Aploactis aspera*)
Velvetfishes (Aploactinidae)
West Pacific, 10 cm

Cockatoo waspfish (*Ablabys taenianotus*)
Waspfishes (Tetrarogidae)
Eastern Indian Ocean and West Pacific, 15 cm

Bandtail waspfish (*Paracentropogon zonatus*)
Waspfishes (Tetrarogidae)
West Pacific, 5 cm

Demon stinger (*Inimicus didactylus*)
Stonefishes (Synanceiidae) Indo-West Pacific, 25 cm
ID: pectoral fins with black central part

Striped stingfish (*Minous trachycephalus*)
Stonefishes (Synanceiidae) Indo-West Pacific, 12 cm
ID: Dark blue iris

Reef stonefish (*Synanceia verrucosa*)
Stonefishes (Synanceiidae)
Indo-Pacific, 40 cm

Giant frogfish (*Antennarius commerson*)
Anglerfishes (Antennariidae)
Indo-Pacific, 45 cm

BONY FISHES — ANGLERFISHES (ANTENNARIIDAE)

Warty frogfish (*Antennarius maculatus*) Anglerfishes (Antennariidae)
Indo-West Pacific, 15 cm
ID: skin is covered with numerous wart-like protuberances

Spotfin frogfish (*Antennarius numifer*)
Anglerfishes (Antennariidae) Indo-Pacific, 13 cm
ID tentative

Painted frogfish (*Antennarius pictus*)
Anglerfishes (Antennariidae) Indo-Pacific, 30 cm
ID: skin is covered by round sponge-like spots

Striated frogfish (*Antennarius striatus*) Anglerfishes (Antennariidae)
Indo-Pacific, Tropical Atlantic, 22 cm. **ID**: skin is covered with filaments resembling hairs, worm-like esca (lure)

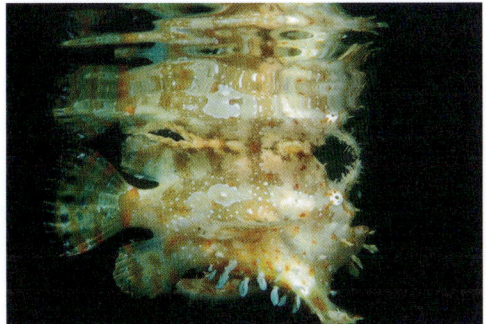 

Sargassumfish (*Histrio histrio*)
Anglerfishes (Antennariidae)
Circumtropical, 20 cm, on Sargassum seaweed

Red-cheeked fairy basslet (*Pseudanthias huchtii*)
Basslets (Anthiinae)
Western Central Pacific, 12 cm

15

BONY FISHES — BASSLETS (ANTHIINAE)

Stocky anthias (*Pseudanthias hypselosoma*)
Basslets (Serranidae, subfamily Anthiinae)
Indo-Pacific, 19 cm

Lori's anthias (*Pseudanthias lori*)
Basslets (Serranidae, subfamily Anthiinae)
Indo-Pacific, 12 cm

Square-spot anthias (*Pseudanthias pleurotaenia*)
Basslets (Serranidae, subfamily Anthiinae)
Indo-West Pacific, 20 cm

Sea goldie (*Pseudanthias squamipinnis*)
Basslets (Serranidae, subfamily Anthiinae)
Indo-West Pacific, 15 cm

Purple anthias (*Pseudanthias tuka*)
Basslets (Serranidae, subfamily Anthiinae)
Indo-West Pacific, 12 cm

Redmouth grouper (*Aethaloperca rogaa*)
Groupers (Serranidae)
Indo-West Pacific, 60 cm

Bluespotted hind (*Cephalopholis cyanostigma*)
Groupers (Serranidae)
West Pacific, 40 cm

Leopard hind (*Cephalopholis leopardus*)
Groupers (Serranidae) Indo-Pacific, 24 cm
ID: upper caudal-fin base with 2 dark brown saddles

BONY FISHES — GROUPERS (SERRANIDAE)

Strawberry hind (*Cephalopholis spiloparaea*)
Groupers (Serranidae)
Indo-Pacific, 30 cm

Darkfin hind (*Cephalopholis urodeta*)
Groupers (Serranidae) Indo-Pacific, 28 cm
ID: pale stripes across lobes of the tail

Coral grouper (*Epinephelus corallicola*)
Groupers (Serranidae)
Western Pacific, 49 cm

Blacktip grouper (*Epinephelus fasciatus*)
Groupers (Serranidae)
Indo-Pacific, 40 cm

Highfin grouper (*Epinephelus maculatus*)
Groupers (Serranidae)
Pacific Ocean, 60,5 cm

Foursaddle grouper (*Epinephelus spilotoceps*)
Groupers (Serranidae)
Indo-West-Pacific, 30 cm

Highfin coralgrouper (*Plectropomus oligocanthus*)
Groupers (Serranidae)
Indo-West Pacific, 75 cm

Yellow-edged lyretail (*Variola louti*)
Groupers (Serranidae) Indo-Pacific, 83 cm
ID: lunate tail with a bright yellow margin

17

BONY FISHES — SOAPFISHES, BIGEYES, GRUNTERS

Barred soapfish (*Diploprion bifasciatum*)
Soapfishes (Serranidae, subfamily Grammistidae)
Indo-West Pacific, 25 cm

Sixstriped soapfish (*Grammistes sexlineatus*)
Soapfishes (Serranidae, subfamily Grammistidae)
Indo-Pacific, 30 cm

Comet longfin (*Calloplesiops altivelis*)
Longfins (Plesiopidae)
Indo-Pacific, 25 cm

Paeony bulleye (*Priacanthus blochii*)
Bigeye (Priacanthidae)
Indo-Pacific, 35 cm

Red bigeye (*Priacanthus macracanthus*)
Bigeyes (Priacanthidae)
West Pacific, 30 cm

Jarbua terapon (*Terapon jarbua*)
Grunters (Terapontidae)
Indo-Pacific, 36 cm

Spotted hawkfish (*Cirrhitichthys aprinus*)
Hawkfishes (Cirrhitidae)
Indo-West Pacific, 12,5 cm

Dwarf hawksfish (*Cirrhitichthys falco*)
Hawkfishes (Cirrhitidae)
Indo-Pacific, 8 cm

BONY FISHES — HAWKFISHES, DOTTYBACKS

Pixie hawkfish (*Cirrhitichthys oxycephalus*)
Hawkfishes (Cirrhitidae)
Indo-Pacific, 10 cm

Longnose hawkfish (*Oxycirrhites typus*)
Hawkfishes (Cirrhitidae)
Indo-Pacific, 13 cm

Diadem dottyback (*Pictichromis diadema*)
Dottybacks (Pseudochromidae)
Western Central Pacific, 6,2 cm

Magenta dottyback (*Pictichromis porphyrea*)
Dottybacks (Pseudochromidae)
West Pacific, 6 cm

Two-lined dottyback (*Pseudochromis bitaeniatus*)
Dottybacks (Pseudochromidae)
Indo-West Pacific, 12 cm

Brown dottyback (*Pseudochromis fuscus*)
Dottybacks (Pseudochromidae)
Indo-Pacific, 10 cm

Jaguar dottyback (*Pseudochromis moorei*) Dottybacks (Pseudochromidae)
Western Central Pacific, 12 cm
Left picture - female, **right picture** - male

BONY FISHES — JAWFISHES, CARDINALFISHES

Bandit dottyback (*Pseudochromis perspicillatus*)
Dottybacks (Pseudochromidae)
West Pacific, 12 cm

Gold-specs jawfish (*Opistognathus randalli*)
Jawfishes (Opistognathidae)
West Pacific, 11 cm

Solor jawfish (*Opistognathus solorensis*)
Jawfishes (Opistognathidae)
Western Pacific, 5 cm

Stripe-finned stalix (*Stalix* sp.)
Jawfishes (Opistognathidae) Philippines, 6 cm
Undescribed species

Black cardinalfish (*Apogonichthyoides melas*)
Cardinalfishes (Apogonidae)
West Pacific, 10 cm

Twinspot cardinalfish (*Taeniamia biguttata*)
Cardinalfishes (Apogonidae)
West Pacific, 11 cm

Dusky-tailed cardinalfish (*Archamia macroptera*)
Cardinalfishes (Apogonidae)
Indo-West Pacific, 9,5 cm

Dog-toothed cardinalfish (*Cheilodipterus isostigmus*)
Cardinalfishes (Apogonidae)
Indo-Pacific, 11 cm

BONY FISHES — CARDINALFISHES (APOGONIDAE)

Half-barred cardinalfish (*Fibramia thermalis*)
Cardinalfishes (Apogonidae)
Indo-Pacific, 9 cm

Flame cardinalfish (*Fowleria flammea*)
Cardinalfishes (Apogonidae)
West Pacific, 3 cm

Marbled cardinalfish (*Fowleria marmorata*)
Cardinalfishes (Apogonidae)
Indo-Pacific, 7,5 cm

Ghost cardinalfish (*Nectamia fusca*)
Cardinalfishes (Apogonidae)
Indo-Pacific, 11 cm

Ring-tailed cardinalfish (*Ostorhinchus aureus*)
Cardinalfishes (Apogonidae)
Indo-West Pacific, 15 cm

Whiteline cardinalfish (*Ostorhinchus cavitensis*)
Cardinalfishes (Apogonidae)
Indo-Pacific, 7,5 cm

Spotted-gill cardinalfish (*Ostorhinchus chrysopomus*) Cardinalfishes (Apogonidae)
Indonesia, Malaysia, Philippines, 10 cm

Yellowlined cardinalfish (*Ostorhinchus chrysotaenia*)
Cardinalfishes (Apogonidae)
Indo-Pacific, 12 cm

BONY FISHES — CARDINALFISHES (APOGONIDAE)

Ochre-striped cardinalfish (*Ostorhinchus compressus*)
Cardinalfishes (Apogonidae)
Indo-West Pacific, 12 cm

Yellowstriped cardinalfish (*Ostorhinchus cyanosoma*)
Cardinalfishes (Apogonidae)
Indo-Pacific, 8 cm

Frostfin cardinalfish (*Ostorhinchus hoevenii*)
Cardinalfishes (Apogonidae)
Indo-West Pacific, 6 cm

Rifle cardinal (*Ostorhinchus kiensis*)
Cardinalfishes (Apogonidae)
Northwest Pacific, 8 cm

Moluccan cardinalfish (*Ostorhinchus moluccensis*)
Cardinalfishes (Apogonidae) Indo-Pacific, 9 cm
ID: white spot behind the the second dorsal fin

Many-lined cardinalfish (*Ostorhinchus multilineatus*)
Cardinalfishes (Apogonidae)
West Pacific, 10 cm

Blackstripe cardinalfish (*Ostorhinchus nigrofasciatus*) Cardinalfishes (Apogonidae)
Indo-Pacific, 10 cm

Rubyspot cardinalfish (*Ostorhinchus rubrimacula*)
Cardinalfishes (Apogonidae)
Western Central Pacific, 4,5 cm

BONY FISHES — CARDINALFISHES (APOGONIDAE)

Seale's cardinalfish (*Ostorhinchus sealei*)
Cardinalfishes (Apogonidae) Indo-Pacific, 10 cm
ID: brownish bars on gill cover

Reef-flat cardinalfish (*Ostorhinchus taeniophorus*)
Cardinalfishes (Apogonidae)
Indo-Pacific, 11,5 cm

Bridled cardinalfish (*Pristiapogon fraenatus*)
Cardinalfishes (Apogonidae)
Indo-Pacific, 11 cm

Iridescent cardinalfish (*Pristiapogon kallopterus*)
Cardinalfishes (Apogonidae)
Indo-Pacific, 15,5 cm

Rufus cardinalfish (*Pristicon rufus*)
Cardinalfishes (Apogonidae)
West Pacific, 11 cm

Jebb's siphonfish (*Siphamia jebbi*)
Cardinalfishes (Apogonidae) Indo-Pacific, 2,5 cm
ID: skin with orange dots

Tubifer cardinalfish (*Siphamia tubifer*)
Cardinalfishes (Apogonidae) Indo-West Pacific, 5 cm, on sea urchins

Pajama cardinalfish (*Sphaeramia nematoptera*)
Cardinalfishes (Apogonidae)
Indo-Pacific, 9 cm

BONY FISHES | GOATFISHES (MULLIDAE)

Orbiculate cardinalfish (*Sphaeramia orbicularis*)
Cardinalfishes (Apogonidae)
Indo-Pacific, 10 cm

Yellowstripe goatfish (*Mulloidichthys flavolineatus*)
Goatfishes (Mullidae)
Indo-Pacific, 43 cm

Yellowfin goatfish (*Mulloidichthys vanicolensis*)
Goatfishes (Mullidae)
Indo-Pacific, 38 cm

Bicolor goatfish (*Parupeneus barberinoides*)
Goatfishes (Mullidae)
West Pacific, 30 cm

Dash-and-dot goatfish (*Parupeneus barberinus*)
Goatfishes (Mullidae)
Indo-Pacific, 60 cm

Manybar goatfish (*Parupeneus multifasciatus*)
Goatfishes (Mullidae)
Pacific Ocean, 35 cm

Freckled goatfish (*Upeneus tragula*)
Goatfishes (Mullidae)
Indo-West Pacific, 25 cm

Blue and gold fusilier (*Caesio caerulaurea*)
Fusiliers (Caesinoidae)
Indo-West Pacific, 35 cm

BONY FISHES — FUSILIERS, EMPERORS

Redbelly yellowtail fusilier (*Caesio cuning*)
Fusiliers (Caesinoidae)
Indo-West Pacific, 60 cm

Yellow-tail fusilier (*Caesio teres*)
Fusiliers (Caesinoidae)
Indo-West Pacific, 40 cm

Banana fusilier (*Pterocaesio pisang*)
Fusiliers (Caesinoidae)
Indo-West Pacific, 21 cm

Randall's fusilier (*Pterocaesio randalli*)
Fusiliers (Caesinoidae)
Indo-West Pacific, 25 cm

Striped large-eye bream (*Gnathodentex aureolineatus*) Emperors (Lethrinidae)
Indo-Pacific, 30 cm

Orange-spotted emperor (*Lethrinus erythracanthus*)
Emperors (Lethrinidae)
Indo-Pacific, 70 cm

Smalltooth emperor (*Lethrinus microdon*)
Emperors (Lethrinidae)
Indo-West Pacific, 80 cm

Orange-striped emperor (*Lethrinus obsoletus*)
Emperors (Lethrinidae) Indo-Pacific, 50 cm.
ID: yellowish stripe from pectoral to caudal fin

BONY FISHES SNAPPERS (LUTJANIDAE)

Ornate emperor (*Lethrinus ornatus*)
Emperors (Lethrinidae)
Indo-West Pacific, 45 cm

Two-spot banded snapper (*Lutjanus biguttatus*)
Snappers (Lutjanidae)
Indo-Pacific, 25 cm

Checkered snapper (*Lutjanus decussatus*)
Snappers (Lutjanidae)
Indo-West Pacific, 35 cm

Humpback red snapper (*Lutjanus gibbus*)
Snappers (Lutjanidae)
Indo-Pacific, 50 cm

Bluestripe snapper (*Lutjanus kasmira*)
Snappers (Lutjanidae) Indo-Pacific, 40 cm
ID: four blue stripes

One-spot snapper (*Lutjanus monostigma*)
Snappers (Lutjanidae)
Indo-Pacific, 60 cm

Emperor red snapper (*Lutjanus sebae*)
Snappers (Lutjanidae)
Indo-West Pacific, 116 cm

Midnight snapper (*Macolor macularis*)
Snappers (Lutjanidae)
West Pacific, 60 cm

BONY FISHES — SPINECHEEKS, SWEETLIPS

Peters' monocle bream (*Scolopsis affinis*)
Spinecheeks (Nemipteridae)
West Pacific, 24 cm

Two-lined monocle bream (*Scolopsis bilineata*)
Spinecheeks (Nemipteridae)
Indo-West Pacific, 25 cm

Saw-jawed monocle bream (*Scolopsis ciliata*)
Spinecheeks (Nemipteridae)
Indo-West Pacific, 25 cm

Pearly monocle bream (*Scolopsis margaritifera*)
Spinecheeks (Nemipteridae) West Pacific, 26 cm.
ID: raws of yellow spots

Oblique-barred monocle bream (*Scolopsis xenochrous*) Spinecheeks (Nemipteridae)
Indo-West Pacific, 22 cm

Lemonfish (*Plectorhinchus flavomaculatus*)
Sweetlips (Haemulidae)
Indo-West Pacific, 60 cm

Harlequin sweetlips (*Plectorhinchus chaetodonoides*) Sweetlips (Haemulidae)
Indo-West Pacific, 75 cm. It is believed that juveniles may imitate toxic flatworms with chaotic movements

BONY FISHES — RUDDERFISHES, TREVALLIES

Yellowbanded sweetlips (*Plectorhinchus lineatus*)
Sweetlips (Haemulidae) West Pacific, 72 cm
ID: diagonal black stripes

Brassy chub (*Kyphosus vaigiensis*)
Rudderfishes (Kyphosidae)
Indo-Pacific, 70 cm

Blue trevally (*Carangoides ferdau*)
Trevallies (Carangidae)
Indo-Pacific, 70 cm

Giant trevally (*Caranx ignobilis*)
Trevallies (Carangidae) Indo-Pacific, 160 cm
ID: small dark spots all-over

Bluefin trevally (*Caranx melampygus*)
Trevallies (Carangidae)
Indo-Pacific, 100 cm

Bigeye trevally (*Caranx sexfasciatus*)
Trevallies (Carangidae)
Indo-Pacific, 85 cm

Golden trevally (*Gnathanodon speciosus*)
Trevallies (Carangidae)
Indo-Pacific, 120 cm

Oxeye scad (*Selar boops*)
Trevallies (Carangidae)
Pacific Ocean, 25 cm

BONY FISHES — SILVERBELLIES, TILEFISHES, REMORAS

Common mojarra (*Gerres oyena*)
Silverbellies (Gerreidae)
Indo-Pacific, 30 cm

Silver moony (*Monodactylus argenteus*)
Moonyfishes (Monodactylidae)
Indo-West Pacific, 13 cm

Stark's tilefish (*Hoplolatilus starcki*)
Tilefishes (Malacanthidae)
Indo Pacific, 15,5 cm

Quakerfish (*Malacanthus brevirostris*)
Tilefishes (Malacanthidae)
Indo-Pacific, 32 cm

Blue blanquillo (*Malacanthus latovittatus*)
Tilefishes (Malacanthidae)
Indo-Pacific, 45 cm

Live sharksucker (*Echeneis naucrates*)
Remoras (Echeneidae)
Circumtropical, 110 cm

Threespot angelfish (*Apolemichthys trimaculatus*)
Angelfishes (Pomacanthidae)
Indo-Pacific, 26 cm

Bicolor angelfish (*Centropyge bicolor*)
Angelfishes (Pomacanthidae)
Indo-Pacific, 15 cm

BONY FISHES — ANGELFISHES (POMACANTHIDAE)

Twospined angelfish (*Centropyge bispinosa*)
Angelfishes (Pomacanthidae)
Indo-Pacific, 10 cm

Barred angelfish (*Paracentropyge multifasciata*)
Angelfishes (Pomacanthidae)
Indo-Pacific, 12 cm

Keyhole angelfish (*Centropyge tibicen*)
Angelfishes (Pomacanthidae)
West Pacific, 19 cm

Pearlscale angelfish (*Centtropyge vroliki*)
Angelfishes (Pomacanthidae)
Indo-West Pacific, 12 cm

Velvet angelfish (*Chaetodontoplus dimidiatus*)
Angelfishes (Pomacanthidae)
Indonesia, 22 cm

Vermiculated angelfish (*Chaetodontoplus mesoleucus*) Angelfishes (Pomacanthidae)
Indo-West Pacific, 18 cm

Emperor angelfish (*Pomacanthus imperator*)
Angelfishes (Pomacanthidae)
Indo-Pacific, 45 cm

Yellowface angelfish (*Pomacanthus xanthometopon*)
Angelfishes (Pomacanthidae)
Indo-Pacific, 38 cm

BONY FISHES — BUTTERFLYFISHES (CHAETODONTIDAE)

Panda butterflyfish (*Chaetodon adiergastos*)
Butterflyfishes (Chaetodontidae)
West Pacific, 20 cm

Sunburst butterflyfish (*Chaetodon kleinii*)
Butterflyfishes (Chaetodontidae)
Indo-Pacific, 15 cm

Raccoon butterflyfish (*Chaetodon lunula*)
Butterflyfishes (Chaetodontidae)
Indo-Pacific, 20 cm

Spotband butterflyfish (*Chaetodon punctatofasciatus*) Butterflyfishes (Chaetodontidae)
Indo-Pacific, 12 cm. **ID**: orange eye bar

Twospot coralfish (*Chaetodon melanopus*)
Butterflyfishes (Chaetodontidae)
West Pacific, 15 cm

Spot-tail butterflyfish (*Chaetodon ocellicaudus*)
Butterflyfishes (Chaetodontidae) Indo-Pacific, 18 cm
ID: round black spot at base of caudal fin

Pacific double-saddle butterflyfish (*Chaetodon ulietensis*) Butterflyfishes (Chaetodontidae)
Indo-Pacific, 15 cm

Copperband butterflyfish (*Chelmon rostratus*)
Butterflyfishes (Chaetodontidae)
West Pacific, 20 cm

BONY FISHES — ANEMONE FISHES (POMACANTHIDAE)

Goldengirdled coralfish (*Coradion chrysozonus*)
Butterflyfishes (Chaetodontidae) Indo-Pacific, 15 cm
ID: dark spot on dorsal fin

Teardrop butterflyfish (*Chaetodon unimaculatus*)
Butterflyfishes (Chaetodontidae)
Indo-Pacific, 20 cm

Longnose butterflyfish (*Forcipiger flavissimus*)
Butterflyfishes (Chaetodontidae)
Indo-Pacific, 22 cm

Pyramid butterflyfish (*Hemitaurichthys polylepis*)
Butterflyfishes (Chaetodontidae)
Pacific Ocean, 18 cm

Yellowtail anemonefish (*Amphiprion clarkii*)
Anemone fishes (Pomacanthidae-Amphiprioninae)
Indo-West Pacific, 15 cm. **ID**: 3 white bars

Tomato anemonefish (*Amphiprion frenatus*)
Anemone fishes (Pomacanthidae-Amphiprioninae)
West Pacific, 14 cm

Clown anemonefish (*Amphiprion ocellaris*)
Anemone fishes (Pomacanthidae-Amphiprioninae)
Indo-West Pacific, 11 cm

Pink anemonefish (*Amphiprion perideraion*)
Anemone fishes (Pomacanthidae-Amphiprioninae)
West Pacific, 10 cm

BONY FISHES — DAMSELFISHES (POMACENTRIDAE)

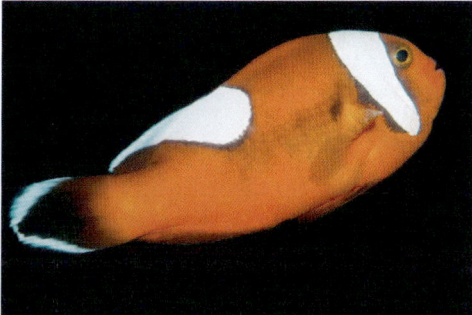

Saddleback anemonefish (*Amphiprion polymnus*) Anemone fishes (Pomacanthidae-Amphiprioninae) West Pacific, 13 cm

Orange skunk anemonefish (*Amphiprion sandaracinos*) Anemone fishes (Pomacanthidae-Amphiprioninae) West Pacific, 14 cm

Spinecheek anemonefish (*Premnas biaculeatus*) Anemone fishes (Pomacanthidae-Amphiprioninae) Indo-West Pacific, 17 cm. **ID**: 3 narrow white bars

Black-tail sergeant (*Abudefduf lorenzi*) Damselfishes (Pomacentridae) Western Central Pacific, 18 cm

Indo-Pacific sergeant (*Abudefduf vaigiensis*) Damselfishes (Pomacentridae) Indo-Pacific, 20 cm **ID**: 5 dark bars

Spiny chromis (*Acanthochromis polyacanthus*) Damselfishes (Pomacentridae) West Pacific, 14 cm

Yellowbelly damselfish (*Amblyglyphidodon leucogaster*) Damselfishes (Pomacentridae) Indo-West Pacific, 13 cm

Ambon chromis (*Chromis amboinensis*) Damselfishes (Pomacentridae) West Pacific, 10 cm. **ID**: orange pectoral fin base

BONY FISHES — DAMSELFISHES (POMACENTRIDAE)

Yellow chromis (*Chromis analis*)
Damselfishes (Pomacentridae)
West Pacific, 17 cm

Dark-fin chromis (*Chromis atripes*)
Damselfishes (Pomacentridae) West Pacific, 10 cm
ID: yellow caudal-fin base with dark margins

Twinspot chromis (*Chromis elerae*)
Damselfishes (Pomacentridae)
Indo-West Pacific, 7 cm

Ovate chromis (*Chromis ovatiformes*)
Damselfishes (Pomacentridae)
West Pacific, 10 cm

Black-bar chromis (*Chromis retrofasciata*)
Damselfishes (Pomacentridae)
West Pacific, 6 cm

Blue-green chromis (*Chromis viridis*)
Damselfishes (Pomacentridae)
Indo-Pacific, 10 cm

Surge damselfish (*Chrysiptera brownriggii*)
Damselfishes (Pomacentridae)
Indo-Pacific, 8 cm

King demoiselle (*Chrysiptera rex*)
Damselfishes (Pomacentridae)
Indo-West Pacific, 7 cm

BONY FISHES — DAMSELFISHES (POMACENTRIDAE)

Rolland's demoiselle (*Chrysiptera rollandi*)
Damselfishes (Pomacentridae) Indo-West Pacific, 7,5 cm
ID: bluish anterior part

Springer's demoiselle (*Chrysiptera* cf. *springeri*)
Damselfishes (Pomacentridae)
West Pacific, 5,5 cm

Talbot's demoiselle (*Chrysiptera talboti*)
Damselfishes (Pomacentridae)
West Pacific, 6 cm

Whitetail dascyllus (*Dascyllus aruanus*)
Damselfishes (Pomacentridae)
Indo-West Pacific, 10 cm

Reticulate dascyllus (*Dascyllus reticulatus*)
Damselfishes (Pomacentridae)
Indo-West Pacific, 9 cm

Threespot dascyllus (*Dascyllus trimaculatus*)
Damselfishes (Pomacentridae)
Indo-Pacific, 14 cm

Black-and-gold chromis (*Neoglyphidodon nigroris*) Damselfishes (Pomacentridae)
Indo-West Pacific, 13 cm. **Left picture**: adult (bicolour form). **Right picture**: juvenile

BONY FISHES — DAMSELFISHES (POMACENTRIDAE)

Barhead damsel (*Neoglyphidodon thoracotaeniatus*) Damselfishes (Pomacentridae)
Western Central Pacific, 13,5 cm. **ID**: 3 dark bars, blue spot on pectoral fin base
Right picture - juvenile

Singlebar devil (*Plectroglyphidodon leucozonus*)
Damselfishes (Pomacentridae)
Indo-Pacific, 12 cm

Alexander's damsel (*Pomacentrus alexanderae*)
Damselfishes (Pomacentridae)
West Pacific, 9 cm

Charcoal damsel (*Pomacentrus brachialis*)
Damselfishes (Pomacentridae)
West Pacific, 8 cm

Neon damselfish (*Pomacentrus coelestis*)
Damselfishes (Pomacentridae)
Indo-West Pacific, 9 cm

Scaly damsel (*Pomacentrus lepidogenys*)
Damselfishes (Pomacentridae)
Indo-West Pacific, 9 cm

Blackmargined damsel (*Pomacentrus nigromarginatus*) Damselfishes (Pomacentridae)
West Pacific, 8 cm

BONY FISHES   PARROTFISHES (SCARIDAE)

Philippine damsel (*Pomacentrus philippinus*)
Damselfishes (Pomacentridae)
Indo-West Pacific, 10 cm

Blueback damsel (*Pomacentrus simsiang*)
Damselfishes (Pomacentridae)
Indo-West Pacific, 7 cm

Ocellate damselfish (*Pomacentrus vaiuli*)
Damselfishes (Pomacentridae)
Pacific Ocean, 10 cm

Daisy parrotfish (*Chlorurus sordidus*)
Parrotfishes (Scaridae)
Indo-Pacific, 40 cm

Yellowfin parrotfish (*Scarus flavipectoralis*)
Parrotfishes (Scaridae) Western Central Pacific, 40 cm
ID: green stripe across head to pectoral fin

Dusky parrotfish (*Scarus niger*)
Parrotfishes (Scaridae) Indo-Pacific, 40 cm
ID: dark-margined bright-green "ear" spot

Rivulated parrotfish (*Scarus rivulatus*)
Parrotfishes (Scaridae)
West Pacific, 40 cm

Tricolour parrotfish (*Scarus tricolor*)
Parrotfishes (Scaridae)
Indo-Pacific, 27 cm

BONY FISHES — WRASSES (LABRIDAE)

Ember parrotfish (*Scarus rubroviolaceus*) Parrotfishes (Scaridae)
Indo-Pacific, 70 cm. **Left picture** - male, **right picture** - female.
ID: semilunar caudal fin (male), reddish with reticulate grey pattern (female)

Tricolour parrotfish (*Scarus tricolor*)
Parrotfishes (Scaridae)
Indo-Pacific, 27 cm

Twospot hogfish (*Bodianus bimaculatus*)
Wrasses (Labridae) Indo-Pacific, 10 cm.
Usually below 40 m

Redfin hogfish (*Bodianus dictynna*)
Wrasses (Labridae)
West Pacific, 15 cm

Splitlevel hogfish (*Bodianus mesothorax*)
Wrasses (Labridae)
West Pacific, 25 cm

Redbreasted wrasse (*Cheilinus fasciatus*)
Wrasses (Labridae)
Indo-Pacific, 40 cm

Tripletail wrasse (*Cheilinus trilobatus*)
Wrasses (Labridae)
Indo-Pacific, 45 cm

BONY FISHES   WRASSES (LABRIDAE)

Humphead wrasse (*Cheilinus undulatus*)
Wrasses (Labridae) Indo-Pacific, 229 cm
ID: thick lips, hump on the forehead

Cigar wrasse (*Cheilio inermis*)
Wrasses (Labridae)
Indo-Pacific, 50 cm

Orange-dotted tuskfish (*Choerodon anchorago*)
Wrasses (Labridae)
Indo-West Pacific, 50 cm

Zoster wrasse (*Choerodon zosterophorus*)
Wrasses (Labridae)
Western Central Pacific, 25 cm

Blueside wrasse (*Cirrhilabrus cyanopleura*)
Wrasses (Labridae)
Eastern Indian Ocean, 15 cm

Lubbock's wrasse (*Cirrhilabrus lubbocki*)
Wrasses (Labridae)
Indonesia, Philippines, 8 cm

Redfin wrasse (*Cirrhilabrus rubripinnis*)
Wrasses (Labridae)
Western Central Pacific, 9 cm

Threadfin wrasse (*Cirrhilabrus temminckii*)
Wrasses (Labridae)
West Pacific, 9 cm

BONY FISHES | WRASSES (LABRIDAE)

Batu coris (*Coris batuensis*)
Wrasses (Labridae) Indo-Pacific, 17 cm
ID: dark bars and white stripes on upper side

African coris (*Coris gaimard*)
Wrasses (Labridae)
Pacific Ocean, 40 cm

Finescale razorfish (*Cymolutes torquatus*)
Wrasses (Labridae)
Indo-Pacific, 20 cm

Yellowtail tubelip (*Diproctacanthus xanthurus*)
Wrasses (Labridae)
Western Central Pacific, 10 cm

Latent slingjaw wrasse (*Epibulus brevis*)
Wrasses (Labridae)
West Pacific, 18,5 cm

Red-lined wrasse (*Halichoeres biocellatus*)
Wrasses (Labridae)
West Pacific, 12 cm

Canary wrasse (*Halichoeres chrysus*)
Wrasses (Labridae)
Eastern Indian Ocean, 12 cm

Hartzfeld's wrasse (*Halichoeres hartzfeldii*)
Wrasses (Labridae)
West Pacific, 18 cm

BONY FISHES   WRASSES (LABRIDAE)

Checkerboard wrasse (*Halichoeres hortulanus*)
Wrasses (Labridae)
Indo-Pacific, 27 cm

Dusky wrasse (*Halichoeres marginatus*)
Indo-Pacific, 18 cm. **ID**: caudal fin with yellow marginal and blue submarginal band

Richmond's wrasse (*Halichoeres richmondi*)
Wrasses (Labridae)
West Pacific, 19 cm

Zigzag wrasse (*Halichoeres scapularis*)
Wrasses (Labridae) Indo-West Pacific, 20 cm
ID: black blotch behind head

Green wrasse (*Halichoeres solorensis*)
Wrasses (Labridae)
West Pacific, 18 cm

Barred thicklip (*Hemigymnus fasciatus*)
Wrasses (Labridae)
Indo-Pacific, 30 cm

Blackeye thicklip (*Hemigymnus melapterus*)
Wrasses (Labridae)
Indo-Pacific, 37 cm

Pastel ringwrasse (*Hologymnosus doliatus*)
Wrasses (Labridae)
Indo-Pacific, 50 cm

41

BONY FISHES — WRASSES (LABRIDAE)

Peacock wrasse (*Iniistius pavo*)
Wrasses (Labridae)
Indo-Pacific, 42 cm

Fivefinger wrasse (*Iniistius pentadactylus*)
Wrasses (Labridae) Indo-Pacific, 25 cm
ID: female with white blotch with red scale margins

Bluestreak cleaner wrasse (*Labroides dimidiatus*)
Wrasses (Labridae)
Indo-Pacific, 14 cm

Northern tubelip (*Labropsis manabei*)
Wrasses (Labridae)
Indo-West Pacific, 10 cm

Rockmover wrasse (*Novaculichthys taeniourus*)
Wrasses (Labridae)
Indo-Pacific, 30 cm

Seagrass wrasse (*Novaculoides macrolepidotus*)
Wrasses (Labridae)
Indo-West Pacific, 16 cm

Two-spot wrasse (*Oxycheilinus bimaculatus*)
Wrasses (Labridae)
Indo-Pacific, 15 cm

Celebes wrasse (*Oxycheilinus celebicus*)
Wrasses (Labridae) West Pacific, 24 cm
ID: dark bar on caudal fin base

BONY FISHES WRASSES (LABRIDAE)

Cockerel wrasse (*Pteragogus enneacanthus*)
Wrasses (Labridae)
West Pacific, 15 cm

Cryptic wrasse (*Pteragogus cryptus*)
Wrasses (Labridae) Indo-West Pacific, 9,5 cm
ID: white stripe above eyes, dark spot on the gill cover

Cocktail wrasse (*Pteragogus flagellifer*)
Wrasses (Labridae)
Indo-West Pacific, 20 cm

Cutribbon wrasse (*Stethojulis interrupta*)
Wrasses (Labridae)
Indo-West Pacific, 13 cm

Three-lined rainbowfish (*Stetojulis trilineata*)
Wrasses (Labridae)
Indo-West Pacific, 15 cm

Sixbar wrasse (*Thalassoma hardwicke*)
Wrasses (Labridae)
Indo-Pacific, 20 cm

Jansens wrasse (*Thalassoma jansenii*)
Wrasses (Labridae)
Indo-West Pacific, 20 cm

Latticed sandperch (*Parapercis clathrata*)
Grubfishes (Pinguiepedidae) Indo-West Pacific, 24 cm.
ID: distinct ocellus on the nape

BONY FISHES   GRUBFISHES (PINGUIEPEDIDAE)

Speckled sandperch (*Parapercis hexophtalma*)
Grubfishes (Pinguiepedidae)
Indo-West Pacific, 29 cm

Nosestripe sandperch (*Parapercis lineopunctata*)
Grubfishes (Pinguiepedidae) West Pacific, 7,4 cm
ID: 2 rows of black dots from eye to middle of body

Redspotted sandperch (*Parapercis schauinslandi*)
Grubfishes (Pinguiepedidae)
Indo-Pacific, 18 cm

Snyder's sandperch (*Parapercis snyderi*)
Grubfishes (Pinguiepedidae)
West Pacific, 11 cm

Yellowbar sandperch (*Parapercis xanthozona*)
Grubfishes (Pinguiepedidae)
Indo-West Pacific, 23 cm

Long-rayed sand-diver (*Trichonotus elegans*)
Sand divers (Trichonotidae)
Indo-West Pacific, 18 cm

Pearly signalfish (*Pteropsaron springeri*) Hemerocoetidae
Western Central Pacific, 3,5 cm. **ID**: males with long filamented dorsal fin, female (right picture) with short and dark dorsal fin

BONY FISHES — BLENNIES (BLENNIDAE)

Marbled stargazer (*Uranoscopus bicinctus*) Stargazers (Uranoscopidae)
Indo-West Pacific, 20 cm. Adult (left picture), sub-adult (right picture)

Convict blenny (*Pholidichthys leucotaenia*)
Convict blennies (Pholidichthyidae) West Pacific,
34 cm. Picture: subadult, 8 cm

Four-fingered lipsucker (*Andamia tetradactylus*)
Blennies (Blenniidae)
West Pacific, 10,5 cm

Bath's comb-tooth (*Ecsenius bathi*)
Blennies (Blenniidae)
Western Central Pacific, 4,4 cm

Twinspot coralblenny (*Ecsenius bimaculatus*)
Blennies (Blenniidae)
West Pacific, 4,5 cm

Forktail blenny (*Meiacanthus atrodorsalis*)
Blennies (Blenniidae)
West Pacific, 11 cm

Striped fang blenny (*Meiacanthus grammistes*)
Blennies (Blenniidae) West Pacific, 11 cm
ID: filamented caudal fin with dark spots

45

BONY FISHES — BLENNIES (BLENNIDAE)

Shorthead fangblenny (*Petroscirtes breviceps*)
Blennies (Blenniidae) Indo-West Pacific, 11 cm
Mimics Meiacanthus species

Floral blenny (*Petroscirtes mitratus*)
Blennies (Blenniidae) Indo-Pacific, 8,5 cm
ID: males with elongated first 3 rays of dorsal fin

Variable sabretooth blenny (*Petroscirtes variabilis*)
Blennies (Blenniidae)
Indo-West Pacific, 15 cm

Bicolour fangblenny (*Plagiotremus laudandus*)
Blennies (Blenniidae)
West Pacific, 8 cm

Jewelled blenny (*Salarias fasciatus*)
Blennies (Blenniidae)
Indo-Pacific, 14 cm

Hairtail blenny (*Xiphasia setifer*)
Blennies (Blenniidae)
Indo-West Pacific, 53 cm

Neglected triplefin (*Helcogramma desa*) 5 cm.
Indonesia, Philippines.

Striped triplefin (*Helcogramma striata*)
Triplefins (Tripterygidae)
West Pacific, 4,3 cm

BONY FISHES

Broadhead clingfish (*Conidens laticephalus*)
Clingfishes (Gobiesocidae) Northwest Pacific, 4 cm
New record for Philippines, inhabits shallow rocks

Oneline clingfish (*Discotrema monogrammum*)
Clingfishes (Gobiesocidae) Indo-West Pacific, 2,4 cm
ID: on crinoids, with white stripe along upper side

Doubleline clingfish (*Lepadichthys lineatus*)
Clingfishes (Gobiesocidae)
Indo-West Pacific, 4,5 cm

Mangrove dragonet (*Callionymus* cf. *enneactis*)
Dragonets (Callionymidae)
West Pacific, 8 cm. ID tentative

Fingered dragonet (*Dactylopus dactylopus*)
Dragonets (Callionymidae) West Pacific, 20 cm
ID: anal fin with blue spots

Orange-black dragonet (*Dactylopus kuiteri*)
Dragonets (Callionymidae) Western Central Pacific,
15 cm. **ID**: dorsal fins black and yellow

Morrison's dragonet (*Synchiropus morrisoni*)
Dragonets (Callionymidae) West Pacific, 8 cm
ID: reddish with scattered white spots. **IT**

Moyer's dragonet (*Synchiropus moyeri*)
Dragonets (Callionymidae)
West Pacific, 7,5 cm

BONY FISHES GOBIES (GOBIIDAE)

Ocellated dragonet (*Synchiropus ocellatus*)
Dragonets (Callionymidae) Pacific Ocean, 9 cm
ID: males with dorsal fin bright orange at the base

Mandarinfish (*Synchiropus splendidus*)
Dragonets (Callionymidae)
West Pacific, 7,5 cm

Diagonal shrimpgoby (*Amblyeleotris diagonalis*)
Gobies (Godiidae)
Indo-Pacific, 11 cm

Giant shrimpgoby (*Amblyeleotris fontanesii*)
Gobies (Godiidae)
West Pacific, 25 cm

Spotted shrimpgoby (*Amblyeleotris guttata*)
Gobies (Godiidae)
West Pacific, 11 cm

Masked shrimpgoby (*Amblyeleotris gymnocephala*)
Gobies (Godiidae) Indo-West Pacific, 14 cm
ID: brown stripe from the eye to gill cover

Masui's shrimpgoby (*Amblyeleotris masuii*)
Gobies (Godiidae)
West Pacific, 11 cm

New-caledonian shrimpgoby (*Amblyeleotris novaecaledoniae*) New Caledonia and PNG, new record for Philippines, 9,5 cm

BONY FISHES — GOBIES (GOBIIDAE)

Periophthalma shrimpgoby (*Amblyeleotris periophthalma*) Gobies (Godiidae)
Indo-West Pacific, 11 cm

Randall's shrimpgoby (*Amblyeleotris randalli*)
Gobies (Godiidae)
West Pacific, 12 cm

Volcano shrimpgoby (*Amblyeleotris rhyax*)
Gobies (Godiidae) West Pacific, 7,3 cm
ID: red bars covered by yellow-orange spots

Redmargin shrimpgoby (*Amblyeleotris rubrimarginata*) Gobies (Godiidae)
Western Central Pacific, 11 cm

Steinitz' shrimpgoby (*Amblyeleotris steinitzi*)
Gobies (Godiidae)
Indo-Pacific, 13 cm

Gorgeous shrimpgoby (*Amblyeleotris wheeleri*)
Gobies (Godiidae)
Indo-Pacific, 10 cm

Flagtail shrimpgoby (*Amblyeleotris yanoi*)
Gobies (Godiidae)
West Pacific, 13 cm

Orange-striped goby (*Amblygobius decussatus*)
Gobies (Godiidae)
Western Central Pacific, 10 cm

BONY FISHES   GOBIES (GOBIIDAE)

Nocturn goby (*Amblygobius nocturnus*)
Gobies (Godiidae) Indo-West Pacific, 10 cm
ID: dark spots under dorsal fin

Whitebarred goby (*Amblygobius phalaena*)
Gobies (Godiidae)
West Pacific Ocean, 15 cm

Threadless cheek-hook goby (*Ancistrogobius yoshigoui*) Gobies (Godiidae) West Pacific, 4 cm
ID: black spot with yellow margin on dorsal fin

Whitespotted frillgoby (*Bathygobius coalitus*)
Gobies (Godiidae)
Indo-Pacific, 10 cm

Cocos frill-goby (*Bathygobius cocosensis*)
Gobies (Godiidae)
Indo-Pacific, 12 cm

Large whip goby (*Bryaninops amplus*)
Gobies (Godiidae)
Indo-Pacific, 4,6 cm

Redeye goby (*Bryaninops natans*)
Gobies (Godiidae)
Indo-Pacific, 2,5 cm

Yellow shrimpgoby (*Cryptocentrus cinctus*)
Gobies (Godiidae) West Pacific, 10 cm.
ID: head and anterodorsal part with bluish spotting

50

BONY FISHES — GOBIES (GOBIIDAE)

Y-bar shrimpgoby (*Cryptocentrus fasciatus*)
Gobies (Godiidae) Indo-West Pacific, 14 cm
ID: no spotting on dorsal fins

Silverspot shrimpgoby (*Ctenogobiops maculosus*) Gobies (Godiidae)
Red Sea, Indo-Pacific, 7 cm

Tangaroa shrimpgoby (*Ctenogobiops tangaroai*)
Gobies (Godiidae)
Pacific Ocean, 6 cm

Blackbelly dwarfgoby (*Eviota atriventris*)
Gobies (Godiidae)
Pacific Ocean, 1,7 cm

Twostripe dwarfgoby (*Eviota bifasciata*)
Gobies (Godiidae)
West Pacific, 3,5 cm

Comet dwarfgoby (*Eviota cometa*)
Gobies (Godiidae) Pacific Ocean, 2,5 cm
ID: dark mark at caudal fin base

Spotted pygmy goby (*Eviota guttata*)
Gobies (Godiidae)
Indo-Pacific, 2,5 cm

Redbelly dwarfgoby (*Eviota nigriventris*)
Gobies (Godiidae)
West Pacific, 3 cm

BONY FISHES GOBIES (GOBIIDAE)

Hairfin dwarfgoby (*Eviota prasites*)
Gobies (Godiidae)
West Pacific, 3 cm

Sebree's pygmy goby (*Eviota sebreei*)
Gobies (Godiidae)
Indo-Pacific, 2,5 cm

Mud reefgoby (*Exyrias belissimus*)
Gobies (Godiidae)
Indo-West Pacific, 15 cm

Innerspotted sandgoby (*Fusigobius inframaculatus*)
Gobies (Godiidae)
Indo-West Pacific, 4,6, cm

Blacktip sandgoby (*Fusigobius melacron*)
Gobies (Godiidae)
Indo-West Pacific, 3,5 cm

Signalfin goby (*Fusigobius signipinnis*)
Gobies (Godiidae)
West Pacific, 5 cm

Eye-bar goby (*Gnatholepis cauerensis*)
Gobionellidae, Indo-Pacific, 6 cm. **ID**: dark eye bar, yellow dark-margined spot above pectoral fin

Yoshino's goby (*Gnatholepis yoshinoi*)
Gobionellidae, West Pacific, 4 cm. **ID**: broad dark bar below eye, yellow spot above pectoral fin

BONY FISHES — GOBIES (GOBIIDAE)

Redstripe goby (*Grallenia rubrilineata*)
Gobies (Godiidae)
Philippines endemic, 1,6 cm

Decorated goby (*Istigobius decoratus*)
Gobies (Godiidae)
Indo-West Pacific, 16 cm

Hector's goby (*Koumansetta hectori*)
Gobies (Godiidae)
Indo-West Pacific, 8,5 cm

Old glory (*Koumansetta rainfordi*)
Gobies (Godiidae)
West Pacific, 8,5 cm

Small goby (*Lubricogobius exiguus*)
Gobies (Godiidae)
West Pacific, 4 cm

Flagfin shrimpgoby (*Mahidolia mystacina*)
Gobies (Gobiidae)
Indo-Pacific, 8 cm

Robust goby (*Oplopomus caninoides*)
Gobies (Godiidae) Indo-West Pacific, 7,5 cm
ID: round brown spots and white spots all over

Spinecheek goby (*Oplopomus oplopomus*)
Gobies (Godiidae) Indo-Pacific, 10 cm
ID: caudal fin with orange stripe

53

BONY FISHES   GOBIES (GOBIIDAE)

Emerald coral goby (*Paragobiodon xanthosoma*) Gobies (Godiidae) Indo-Pacific, 4 cm, on Seriatopora branching coral

Caroline Islands ghost goby (*Pleurosicya carolinensis*) Gobies (Godiidae) Philippines, Central Pacific, 2,5 cm

Bubble coral goby (*Pleurosicya micheli*) Gobies (Godiidae) Indo-Pacific, 3 cm, on hard corals

Toothy goby (*Pleurosicya mossambica*) Gobies (Godiidae) Red Sea, Indo-Pacific, 2 cm

Banded reefgoby (*Priolepis cinctus*) Gobies (Godiidae) Indo-West Pacific, 5 cm

Orange reefgoby (*Priolepis nuchifasciata*) Gobies (Godiidae) West Pacific, 4 cm

Lanceolate shrimpgoby (*Tomiyamichthys lanceolatus*) Gobies (Godiidae) West Pacific, 11 cm. **ID**: dark spots on body and dorsal fin

Magnificient shrimpgoby (*Tomiyamichthys* sp.) Gobies (Godiidae) West Pacific, 6 cm

BONY FISHES — GOBIES (GOBIIDAE)

Ring-eye dwarfgoby (*Trimma benjamini*)
Gobies (Godiidae)
West Pacific, 3 cm

Candycane dwarfgoby (*Trimma cana*)
Gobies (Godiidae)
West Pacific, 2,5 cm

Blue-striped cave goby (*Trimma caudomaculatum*)
Gobies (Godiidae)
West Pacific, 4,5 cm

Flame goby (*Trimma macrophthalmus*)
Gobies (Godiidae)
Indo-West Pacific, 2,5 cm

Red-spotted dwarfgoby (*Trimma rubromaculatum*)
Gobies (Godiidae)
West Pacific, 3,5 cm

Twostripe goby (*Valenciennea helsdingenii*)
Gobies (Godiidae)
Indo-West Pacific, 25 cm

Orange spotted sleeper goby (*Valenciennea puellaris*) Gobies (Gobiidae)
Indo-Pacific, 22 cm

Greenband goby (*Valenciennea randalli*)
Gobies (Godiidae)
West Pacific, 12 cm

BONY FISHES — GOBIES (GOBIIDAE)

Sixspot goby (*Valenciennea sexguttata*)
Gobies (Gobiidae) Indo-Pacific, 14 cm. **ID**: silvery spots on cheeks, first dorsal fin with black tip

Ambanoro shrimpgoby (*Vanderhorstia ambanoro*)
Gobies (Gobiidae)
Indo-West Pacific, 13 cm

Gold-marked shrimpgoby (*Vanderhorstia auronotata*)
Gobies (Gobiidae) Indonesia, Philippines, 5 cm
ID: vertical yellow lines and dark bars

Dorsalspot shrimpgoby (*Vanderhorstia dorsomacula*)
Gobies (Gobiidae)
West Pacific, 5 cm

Butterfly shrimpgoby (*Vanderhorstia papilio*)
Gobies (Gobiidae) Indo-Pacific, 4,1 cm
ID: numerous yellow dark-margined spots

Blueshine shrimpgoby (*Vanderhorstia* sp.)
Gobies (Gobiidae)
West Pacific, 10 cm

Elegant firefish (*Nemateleotris decora*)
Dart gobies (Ptereleotridae)
Indo-Pacific, 9 cm

Fire goby (*Nemateleotris magnifica*)
Dart gobies (Ptereleotridae)
Indo-Pacific, 9 cm

BONY FISHES / BATFISHES (EPHIPPIDAE)

Lowfin dartfish (*Ptereleotris brachyptera*) Dart gobies (Ptereleotridae) Western Central Pacific, 5,4 cm. **ID**: pectoral fins with a narrow red bar at base

Blackfin dartfish (*Ptereleotris evides*) Dart gobies (Ptereleotridae) Indo-Pacific, 14 cm

Curious wormfish (*Gunnellichthys curiosus*) Worm gobies (Microdesmidae) Indo-Pacific, 11 cm

Yellowstripe wormfish (*Gunnellichthys viridescens*) Worm gobies (Microdesmidae) Indo-Pacific, 9 cm

Golden spadefish (*Platax boersii*) Batfishes (Ephippidae) Indo-West Pacific, 40 cm **ID**: rounded head, anal fin with black margin

Orbicular batfish (*Platax orbicularis*) Batfishes (Ephippidae) Indo-Pacific, 60 cm. **ID**: brown eye bar

Dusky batfish (*Platax pinnatus*) Batfishes (Ephippidae) West Pacific, 45 cm

Blanthead batfish (*Platax teira*) Batfishes (Ephippidae) Indo-West Pacific, 70 cm

BONY FISHES — RABBITFISHES (SIGANIDAE)

Mottled spinefoot (*Siganus fuscescens*)
Rabbitfishes (Siganidae)
West Pacific, 40 cm

Golden-lined spinefoot (*Siganus lineatus*)
Rabbitfishes (Siganidae)
Indo-West Pacific, 43 cm

Masked spinefoot (*Siganus puellus*)
Rabbitfishes (Siganidae)
Indo-West Pacific, 38 cm

Little spinefoot (*Siganus spinus*)
Rabbitfishes (Siganidae)
Indo-West Pacific, 28 cm

Barhead spinefoot (*Siganus virgatus*)
Rabbitfishes (Siganidae)
Indo-West Pacific, 30 cm

Convict surgeonfish (*Acanthurus triostegus*)
Surgeonfishes (Acanthuridae)
Indo-Pacific, 27 cm

Chocolate surgeonfish (*Acanthurus pyroferus*) Surgeonfishes (Acanthuridae)
Indo-Pacific, 29 cm. **ID**: orange blotch around pectoral fin base. Juvenile (left picture) mimics **Pearlscale angelfish** (*Centropyge vrolikii*)

BONY FISHES   SURGEONFISHES (ACANTHURIDAE)

Twospot surgeonfish (*Ctenochaetus binotatus*)
Surgeonfishes (Acanthuridae) Indo-Pacific, 22 cm.
ID: blue ring around eye

Slender unicorn (*Naso minor*)
Surgeonfishes (Acanthuridae)
Indo-West Pacific, 30 cm

Bluespine unicornfish (*Naso unicornis*)
Surgeonfishes (Acanthuridae)
Indo-Pacific, 70 cm

Sailfin tang (*Zebrasoma velifer*)
Surgeonfishes (Acanthuridae)
West Pacific, 40 cm

Moorish idol (*Zanclus cornutus*)
Moorish idols (Zanclidae)
Indo-Pacific, 23 cm

Obtuse barracuda (*Sphyraena obtusata*)
Barracudas (Sphyraenidae)
Indo-Pacific, 55 cm

Bali sardinella (*Sardinella lemuru*) Clupeiformes (Herrings)
Indo-West Pacific, 23 cm. Famous "Moalboal Surdine Run" is a popular attraction for divers and snorkelers where millions of fishes congregate right at the reef drop-off on Panagsama beach.

BONY FISHES — SOLES (SOLEIDAE)

Leopard flounder (*Bothus pantherinus*)
Left-eyed flounders (Bothidae) Indo-Pacific, 39 cm.
ID: numerous flower-like blotches

Dwarf sole (*Aseraggodes xenicus*)
Soles (Soleidae)
Indo-West Pacific, 8 cm

White sole (*Aseragoddes dubius*)
Soles (Soleidae)
West Pacific, 8 cm. ID tentative

Small sole (*Aseragoddes* sp.)
Soles (Soleidae)
Indonesia, Philippines, 2,5 cm

Hook-nosed sole (*Heteromycteris hartzfeldii*)
Soles (Soleidae)
West Pacific, 20 cm

Carpet sole (*Liachirus melanospilos*)
Soles (Soleidae) Indo-West Pacific, 15 cm
ID: black dots, brown spots

Peacock sole (*Pardachirus pavoninus*)
Soles (Soleidae)
Indo-Pacific, 25 cm

Black-tip sole (*Soleichthys heterorhinos*)
Soles (Soleidae) Indo-West Pacific, 18 cm.
ID: tubular nostril ends above eyes

BONY FISHES   TRIGGERFISHES (BALLISTIDAE)

Yellow-spot-margined sole (*Soleichthys* sp.)
Soles (Soleidae)
West Pacific, 1 cm

Zebra sole (*Zebrias zebra*)
Soles (Soleidae) West Pacific, 19 cm.
ID: black caudal fin with yellow spots

Shorteheaded tonguesole (*Cynoglossus kopsii*)
Tongue soles (Cynoglossidae)
Indo-West Pacific, 60 cm

Clown triggerfish (*Balistoides conspicillum*)
Triggerfishes (Ballistidae)
Indo-Pacific, 50 cm

Titan triggerfish (*Balistoides viridescens*)
Triggerfishes (Ballistidae)
Indo-West Pacific, 75 cm

Red-toothed triggerfish (*Odonus niger*)
Triggerfishes (Ballistidae)
Indo-Pacific, 50 cm

Blue triggerfish (*Pseudobalistes fuscus*)
Triggerfishes (Ballistidae) Indo-Pacific, 55 cm.
ID: subadults with golden spots, joined in lines

Boomerang triggerfish (*Sufflamen bursa*)
Triggerfishes (Ballistidae)
Indo-Pacific, 25 cm

BONY FISHES — FILEFISHES (MONACANTHIDAE)

Halfmoon triggerfish (*Sufflamen chrysopterum*)
Triggerfishes (Ballistidae)
Indo-West Pacific, 30 cm

Radial leatherjacket (*Acreichthys radiatus*)
Filefishes (Monacanthidae) West Pacific, 7 cm.
Near xeniid corals

Gilded triggerfish (*Xanthichthys auromarginatus*) Triggerfishes (Ballistidae)
Indo-Pacific, 30 cm. **ID**: Males (left picture): blue patch across the jaw and throat. Females (right picture): caudal fin with red margin

Bristle-tail filefish (*Acreichthys tomentosus*)
Filefishes (Monacanthidae)
Indo-West Pacific, 12 cm

Scribbled leatherjacket filefish (*Aluterus scriptus*)
Filefishes (Monacanthidae)
Circumtropical, 110 cm

Broom filefish (*Amanses scopas*)
Filefishes (Monacanthidae)
Indo-Pacific, 20 cm

Spectacled filefish (*Cantherhines fronticinctus*)
Filefishes (Monacanthidae)
Indo-West Pacific, 25 cm

BONY FISHES — FILEFISHES (MONACANTHIDAE)

Honeycomb filefish (*Cantherhines pardalis*)
Filefishes (Monacanthidae)
Indo-Pacific, 25 cm

Harlequin filefish (*Oxymonacanthus longirostris*)
Filefishes (Monacanthidae)
Indo-Pacific, 12 cm

False puffer (*Paraluteres prionurus*)
Filefishes (Monacanthidae) Indo-Pacific, 11 cm.
Mimics toxic puffer *Canthigaster valentini*

Pig faced leather jacket (*Paramonacanthus choirocephalus*) Filefishes (Monacanthidae)
West Pacific, 11 cm

Shortsnout filefish (*Paramonacanthus curtorhynchos*) Filefishes (Monacanthidae)
Eastern Indian Ocean and West Pacific, 11 cm

Redtail filefish (*Pervagor melanocephalus*)
Filefishes (Monacanthidae) Indo-West Pacific, 16 cm

Rhinoceros leatherjacket (*Pseudalutarius nasicornis*) Filefishes (Monacanthidae)
Indo-West Pacific, 19 cm

Strap-weed filefish (*Pseudomonacanthus macrurus*) Filefishes (Monacanthidae)
Indo-West Pacific, 18 cm

BONY FISHES — FILEFISHES (MONACANTHIDAE)

Diamond leatherjacket (*Rudarius excelsus*)
Filefishes (Monacanthidae)
West Pacific, 2,5 cm

Longhorn cowfish (*Lactoria cornuta*)
Boxfishes (Ostraciidae)
Indo-Pacific, 46 cm

Reticulate boxfish (*Ostracion solorensis*) Boxfishes (Ostraciidae)
West Pacific, 12 cm. **Left picture** - female, **right** - male

Yellow boxfish (*Ostracion cubicus*)
Boxfishes (Ostraciidae) Indo-Pacific, 45 cm.
ID: yellow net on body, yellowish caudal fin base

Whitespotted boxfish (*Ostrasion meleagris*)
Boxfishes (Ostraciidae) Indo-Pacific, 25 cm.
ID: male with yellow stripe along carapace top

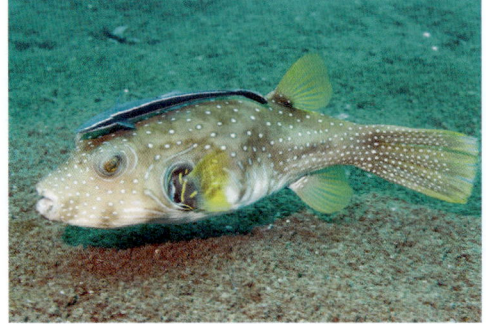

Whitespotted puffer (*Arothron hispidus*)
Pufferfishes (Tetraodontidae)
Indo-Pacific, 55 cm

Immaculate puffer (*Arothron immaculatus*)
Pufferfishes (Tetraodontidae)
Indo-West Pacific, 30 cm

BONY FISHES — PUFFERFISHES (TETRAODONTIDAE)

Narrow-lined puffer (*Arothron manilensis*)
Pufferfishes (Tetraodontidae)
West Pacific, 31 cm

Black-spotted puffer (*Arothron nigropunctatus*)
Pufferfishes (Tetraodontidae)
Indo-Pacific, 33 cm

Starry puffer (*Arothron stellatus*)
Pufferfishes (Tetraodontidae) Indo-Pacific, 120 cm.
ID: small dark spots all over

Compressed toby (*Canthigaster compressa*)
Pufferfishes (Tetraodontidae)
West Pacific, 12 cm

Papuan toby (*Canthigaster papua*)
Pufferfishes (Tetraodontidae)
Indo-West Pacific, 10 cm

Blacksaddle toby (*Canthigaster valentini*)
Pufferfishes (Tetraodontidae)
Indo-Pacific, 11 cm

Birdbeak burrfish (*Cyclichthys orbicularis*)
Porcupinefishes (Diodontidae)
Indo-West Pacific, 15 cm

Longspined porcupinefish (*Diodon holocanthus*)
Porcupinefishes (Diodontidae) Circumtropical, 50 cm.
ID: wide brown eye bar

SHARKS

Whale shark (*Rhincodon typus*) Whale sharks (Rhincodontidae) Circumglobal. Size: up to 20 m (females), 16 m (males), rarely above 12 m. Feeds on plankton, small fish, squids.
Pictures: **Alex Stoyda**

Pelagic thresher shark (*Alopias pelagicus*) Thresher sharks (Alopiidae)
Indo-Pacific. Size and weight: 3,3 m, 88 kg. Feeds on small pelagic prey.
ID: dark pigment above the rounded pectoral fins. Pictures: **Alex Stoyda.** Monad Shoal, Malapascua Island

Whitetip reef shark (*Triaenodon obesus*)
Requiem sharks (Carcharhinidae)
Indo-Pacific, 2,1 m

Blacktail reef shark (*Carcharhinus amblyrhynchos*)
Requiem sharks (Carcharhinidae)
Indo-Pacific, 2,5 m

Tawny nurse shark (*Nebrius ferrugineus*)
Nurse sharks (Ginglymostomatidae)
Indo-Pacific, 3,2 m

Zebra shark (*Stegostoma fasciatum*)
Zebra sharks (Stegostomatidae)
Indo-Pacific, 3,5 m

RAYS, SNAKES, TURTLES

Blue-spotted stingray (*Dasyatis kuhlii*)
Stingrays (Dasyatidae)
Indo-Pacific, 70 cm

Giant manta (*Manta birostris*)
Eagle and Manta rays (Myliobatidae)
Circumglobal, 7 m (disc size)

Ribbontail stingray (*Taeniura lymma*)
Stingrays (Dasyatidae)
Indo-Pacific, 40 cm

Turtle-headed sea snake (*Emydocephalus annulatus*)
Elapidae, Western and Central Pacific, 103 cm
ID: zigzag stripes

Banded sea krait (*Laticauda colubrina*)
Elapidae, Indo-West Pacific, 1.4 m (females), 0.9 m (males) **ID**: yellowish snout, black stripes

Green sea turtle (*Chelonia mydas*)
Sea turtles (Chelonidae)
Circumtropical, 1,5 m

Hawksbill turtle (*Eretmochelys imbricata*)
Sea turtles (Chelonidae) Circumtropical, 1 m. Feeds on sea sponges, ascidians, cnidarians, jellyfish.
ID: beak-like mouth, carapace's serrated margin, two pairs of scales between the eyes

SHRIMPS — COMMENSAL SHRIMPS (PALAEMONIDAE)

Twin-stripe crinoid shrimp (*Periclimenes* cf. *affinis*)
Indo-Pacific, 2 cm.
ID: 2 longitudinal white stripes

Ambon crinoid shrimp (*Laomenes amboinensis*)
Indo-West Pacific, 2 cm.
ID: presence of rostral teeth

Leopard crinoid shrimp (*Laomenes* cf. *pardus*)
Indo-Pacific, 2,0 cm. **ID**: close to *L.pardus*, different host and colour pattern

Leopard crinoid shrimp (*Laomenes pardus*)
Indo-Pacific, 1,8 cm
ID: white with transverse dark bands

White-top crinoid shrimp (*Laomenes* sp.)
Philippines, 1,8 cm. **ID:** broad longitudinal white stripe on the top of the body

Polka dots crinoid shrimp (*Laomenes* sp.)
Philippines, 2 cm.
ID: large round dots repeated in even rows

Donald Duck shrimp (*Leander plumosus*)
Indo-West Pacific, 3 cm
ID: long flattened rostrum with sensory hairs

Brook's urchin shrimp (*Allopontonia brooki*)
Indo-Pacific. On fire urchins, 2-2,2 cm
ID: distinctive coloration, smooth chelipeds

SHRIMPS COMMENSAL SHRIMPS (PALAEMONIDAE)

Coleman's shrimp (*Periclimenes colemani*)
Indo-Pacific, 2 cm.
On *Asthenosoma* fire urchins

Hairy urchin shrimp (*Sandimenes hirsutus*)
Indo-Pacific, 2,2 cm. On *Asthenosoma* fire urchins.
ID: longitudinal white stripe

Purple urchin shrimp (*Stegopontonia commensalis*)
Indo-Pacific, 3 cm, sea urchins *Echinothrix* sp.
ID: broad longitudinal white stripe

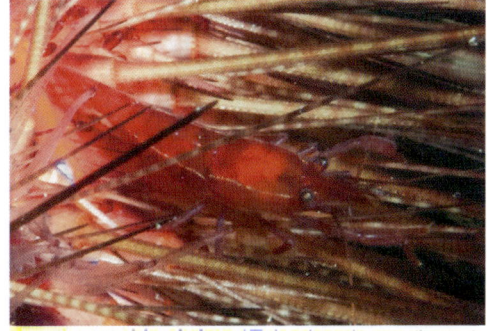

Zanzibar urchin shrimp (*Tuleariocaris zanzibarica*)
Indo-Pacific, on sea urchins *Astropyga*.
ID: black eyes, 2 thin white stripes

Emperor shrimp (*Zenopontonia rex*) Indo-Pacific.
On echinoderms, 2-3 cm. **ID:** coloration variable,
rostrum looks smooth due to tiny spines

Sea star shrimp (*Zenopontonia soror*)
Indo-Pacific. On echinoderms, 1 cm.
ID: small shrimp with dentate rostrum (10-11 teeth)

Red-dotted ancylomenes (*Ancylomenes adularans*)
West Pacific, 2,5 cm. **ID:** red spots over body,
bigger spots before tail base and white saddle

Holthus' anemone shrimp (*Ancylomenes holthuisi*)
Indo-West Pacific, 3 cm. **ID:** brown border around
white saddle and tail base

SHRIMPS   COMMENSAL SHRIMPS (PALAEMONIDAE)

Yellow-spotted anemone shrimp (*Ancylomenes luteomaculatus*) West Pacific, 3 cm.
ID: white saddle with red marking

Magnificent anemone shrimp (*Ancylomenes magnificus*) West Pacific, 3,5 cm.
ID: saddle and curled inward claws are white

Sarasvati anemone shrimp (*Ancylomenes sarasvati*) West Pacific, 2,5 cm
ID: bright blue border before the white saddle

Graceful anemone shrimp (*Ancylomenes venustus*) West Pacific, 3 cm. **ID:** white-speckled claw arms, round saddle with bluish sub-marginal band

Egg shell shrimp (*Hamopontonia fungicola*) Indo-West Pacific, 2,5 cm, on *Heliofungia* corals
ID: 2 large white saddles, tiny reddish dots

Leopard anemone shrimp (*Izucaris masudai*) I-W-P, 2 cm. Associated with sea anemone *Nemanthus* sp. Coloration matches host anemone. Photo: **Alexis Pricipe**

Peacock-tail shrimp (*Ancylocaris brevicarpalis*) Indo-Pacific, 4 cm, on sea anemones.
ID: tail with 5 large purple spots with yellow centres

Ornate coral shrimp (*Actinimenes inornatus*) Indo-Pacific, 2 cm, anemone-associated. **ID:** broad white stripes between eyes, from eyes to tail base

SHRIMPS   COMMENSAL SHRIMPS (PALAEMONIDAE)

Chocolate shrimp (*Phycomenes* sp.) Philippines, Indonesia, 1 cm. Inhabits zoanthids on worm tubes. **ID:** brown with white transverse banding. ID tentative

Hidden corallimorph shrimp (*Pliopontonia furtiva*) Indo-West Pacific, 2,5 cm, on corallimorph *Rhodactis*. **ID:** transparent body, white bands, white & yellow spots

Maldivian pontonides (*Pontonides maldivensis*) Indo-Pacific, 2 cm, on *Tubastraea* corals. **ID:** recognizable orange coloration

Anchor coral shrimp (*Vir euphyllius*) West Pacific, 1,5 cm, on Anchor coral *Euphyllia* sp. **ID:** white blotch behind eyes, red dots on legs and body

Bubble coral shrimp (*Vir philippinensis*) Indo-West Pacific, 1,5 cm. **ID:** purple lines on legs and claw arms

Sea pen shrimp (*Dasycaris ceratops*) Indo-West Pacific, 1,5 cm, on sea pens **ID**: 5 rostrum teeth, fine wavy red lines on carapace

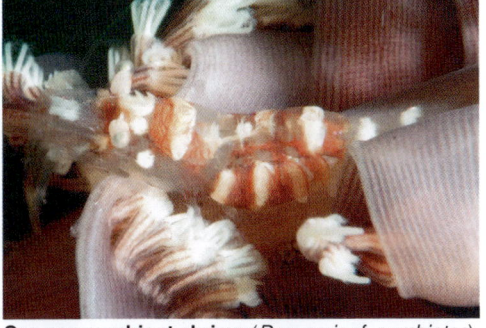

Sea pen symbiont shrimp (*Dasycaris* cf. *symbiotes*) West Pacific, 1, 5 cm, on sea pens. **ID:** globular eyes (ovoid cornea in *D.ceratops*)

Zanzibar whip coral shrimp (*Dasycaris zanzibarica*) Indo-Pacific, 1,4 cm. **ID**: toothless rostrum, white transverse banding

SHRIMPS — COMMENSAL SHRIMPS (PALAEMONIDAE)

Bandaged gorgonian shrimp (*Hamodactylus* sp.) Indonesia, 1 cm.
ID: probably *H.boschmai*, less acute rostrum tip

Translucent gorgonian shrimp (*Manipontonia psamathe*) Indo-Pacific, 2,5 cm, on gorgonians, black corals. **ID**: rostrum with 6 teeth, 2 red lines

Yonge's gorgonian shrimp (*Miopontonia yongei*) Indo-West Pacific, 1,2 cm.
ID: translucent, often covered by golden spots

Anker's wire coral shrimp (*Pontonides ankeri*) Indo-Pacific, 1,5 cm, on wire corals. Several similar species. Probably all of them are forms of *P.loloata*

Maldivian red-eyed shrimp (*Exoclimenella maldivensis*) Indo-West Pacific, 3 cm
ID: red eyes, white-dotted claw arms

Pacific palaemon (*Palaemon pacificus*) Indo-Pacific, 3 cm. Shallow boulders zone.
ID: translucent, thin brown transverse banding

Red-stripe gorgonian shrimp (*Rapipontonia galene*) West Pacific, 2 cm. **ID**: covered with longitudinal bands of black chromatophores

Hydroid shrimp (*Rapipontonia hydra*) West Pacific, 1 cm, on plumularid hydroids
ID: longitudinal yellowish stripes

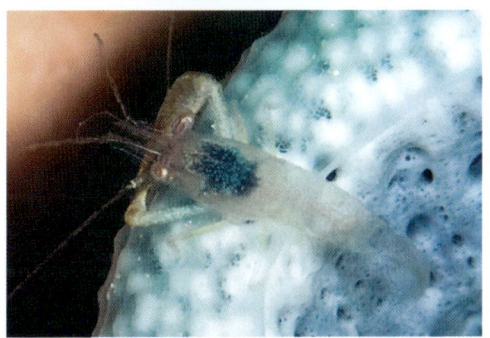

SHRIMPS — COMMENSAL SHRIMPS (PALAEMONIDAE)

Commensal sponge shrimp (*Thaumastocaris streptopus*) Indo-Pacific, 1,5 cm, inside tube sponges. **ID**: translucent, dark eyes

Clear cleaner shrimp (*Urocaridella antonbruunii*) Indo-Pacific, 2,5 cm. **ID**: rostrum with red and white bands, translucent at base

Golden cleaner shrimp (*Urocaridella* sp.) West Pacific, 2,5 cm. **ID**: readily identified by golden lines on rostrum, claw arms and abdomen

Speckled shrimp (*Urocaridella urocaridella*) Indo-West Pacific, 3 cm. **ID**: small yellow spots, forming transverse banding

Flattened shrimp (*Harpiliopsis depressa*) Indo-Pacific, 1,2 cm, on branching corals. **ID**: flattened carapace with fine brown lines

Cousin coral shrimp (*Harpilius consobrinus*) Indo-Pacific, 1,8 cm, on branching corals. **ID**: Claw arms and body covered by small golden spots

Golden-dotted ascidian shrimp (*Dactylonia* sp.) Philippines, 1 cm, on ascidians. **ID**: translucent with small golden spots, minor P2 finger short

Red-dotted tunicate shrimp (*Odontonia rufopunctata*) Indo-West Pacific, 0,6 cm, on ascidians **ID**: scattered red dots

SHRIMPS — COMMENSAL SHRIMPS (PALAEMONIDAE)

Green ascidian shrimp (*Periclimenaeus storchi*) West Pacific, 1 cm, on ascidians **ID**: translucent greenish, covered with white-dots

Mushroom coral shrimp (*Cuapetes kororensis*) West Pacific, 4 cm. **ID**: reddish carapace, translucent claws and abdomen, white head

Spotted-arm cuapetes shrimp (*Cuapetes lacertae*) West Pacific, 3,5 cm. **ID:** brown-spotted claw arms

Red-stripe cuapetes (*Cuapetes* sp.) Philippines, 2,8 cm. **ID**: several slanted red lines on lower abdomen

Red claw cuapetes shrimp (*Cuapetes tenuipes*) Indo-West Pacific, 3,5 cm. **ID**: orange claws, red spots on claw arms, three red longitudinal body stripes

Tiger snapping shrimp (*Alpheus bellulus*) Indo-West Pacific, 4 cm, goby associate. **ID**: brown-orange net pattern. Picture- **I.Khlopunova**

Big snapping shrimp (*Alpheus digitalis* species complex), West Pacific, 6 cm, in association with *V.ambanoro* goby. **ID**: bluish legs

Djedda alpheus (*Alpheus djeddensis*) Indo-Pacific, 5 cm, goby associate **ID**: pale, dark-red blotch on carapace

SHRIMPS   SNAPPING SHRIMPS (ALPHEIDAE)

Randall's snapping shrimp (*Alpheus randalli*)
Indo-Pacific, 3 cm, goby associate
ID: yellow legs, red-white banding

Two ribbed alpheus (*Alpheus bicostatus*)
Indo-West Pacific, 3 cm.
ID: prominent eyes, brown with white spots

Red snapping shrimp (*Alpheus bisincisus*)
Indo-West Pacific, 4 cm.
ID: opaque red, small dark spots

Small alpheus (*Alpheus* cf. b*revipes, A. critnitus* group) Philippines, PNG, 2 cm
ID: white dark-bordered eyes

White-dotted snapping shrimp (*Alpheus* sp., *A. crinitus* group) Philippines, 2,5 cm
ID: opaque reddish-green with small white spots

Coral snapping shrimp (*Alpheus lottini*) Indo-Pacific, 4 cm, on Pocillopora branching corals. **ID**: orange with dark red spots on arm claws and carapace

Greenish snapping shrimp (*Alpheus pacificus* species complex) Philippines, 3 cm.
ID: dark brown with fine green pattern and white dots

Ocellated snapping shrimp (*Alpheus* sp.)
Philippines, 3 cm
ID: black yellow-bordered ocelli on abdomen

SHRIMPS — BROKEN BACK SHRIMPS (HIPPOLYTIDAE)

White-dotted snapping shrimp (*Alpheus* sp., *A. crinitus* group) Philippines, 2,5 cm
ID: opaque reddish-green with small white spots

Djibouti snapping shrimp (*Athanas djiboutensis*) Indo-Pacific, 1 cm, nocturnal. ID: brown with broad white central stripe from rostrum to tail

Deman's snapping shrimp (*Synalpheus demani*) Indo-Pacific, 3 cm. On crinoids, usually in pairs.
ID: dark reddish-brown

Soft coral snapping shrimp (*Synalpheus neomeris*) Indo-West Pacific, 2,8 cm, on soft corals
ID: translucent, milky-white spots

Xenia shrimp (*Alcyonohippolyte commensalis*) West Pacific, 1 cm, on *Xenia* soft corals
ID: white with reticulating olive-brown pattern

Humpback soft coral shrimp (*Alcyonohippolyte dossena*) Indo-West Pacific, 1 cm
ID: mimics soft coral polyps by swellings

Zoanthid shrimp (*Alcyonohippolyte* sp.) West Pacific, 1 cm, on zoanthids
ID: brown with white spots, mimicking host

Two-stripe soft coral shrimp (*Alcyonohippolyte tenuicarpus*) West Pacific, 1 cm
ID: two broad reticulated olive stripes

SHRIMPS — BROKEN BACK SHRIMPS (HIPPOLYTIDAE)

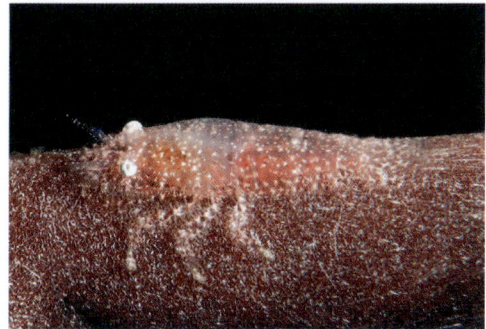

Cryptic sponge shrimp (*Gelastocaris paronae*)
Indo-Pacific, 2 cm. **ID**: Flattened body, perfectly mimics variable host sponge colours and patterns

Broken-back soft coral shrimp (*Hippolyte* sp.)
Philippines, Indonesia, 1 cm, on soft corals.
ID: smooth translucent body, mimics soft coral host

Hydroid white-stripe shrimp (*Hippolyte* sp.)
Philippines, Indonesia, 1 cm, on hydroids
ID: central white stripe

Hump-back cleaner shrimp (*Lysmata amboinensis*)
Indo-Pacific, 3 cm. **ID**: orange, central white stripe, bordered by two red stripes

Striped small cleaner (*Lysmatella prima*)
Indo-West Pacific, 2,5 cm. **ID:** yellow with longitudinal white stripes, yellow legs and white antennae

Red algae shrimp (*Phycocaris simulans*)
West Pacific, 1 cm.
ID: tiny, mimics floating red algae

Green hairy shrimp (*Phycocaris* cf. s*imulans*)
West Pacific, 1 cm. **ID**: semi-translucent, greenish, near green ascidians

Green marbled shrimp (*Saron* cf. *marmoratus*)
Indonesia, 3,5 cm
ID: turquoise irregular spots, dark green tail

SHRIMPS   BROKEN BACK SHRIMPS (HIPPOLYTIDAE)

Marbled shrimp (round-spotted) (*Saron* cf. *marmoratus*) Maldives, Indonesia, Philippines, 6 cm
ID: round spots on carapace

Marbled shrimp (starry) (*Saron* cf. *marmoratus*) Philippines, PNG, 6 cm
ID: red stars on pink carapace and abdomen

Marbled shrimp (islands) (*Saron* cf. *marmoratus*) West Pacific, 4,5 cm. **ID**: irregular brown or red blue-dotted spots on pink background

Marbled shrimp (fine tracks) (*Saron* cf. *marmoratus*) Philippines, 5 cm.
ID: fine red lines between irregular red spots

Marble shrimp (cells) (*Saron* cf. *marmoratus*) Philippines, 4,4 cm
ID: big round spots with small white spots inside

Marbled shrimp (*Saron marmoratus*) Indo-Pacific, 4 cm.
ID: pink with reddish blue-bordered round spots. **IT**

Coral marbled shrimp (*Saron neglectus*) Indo-Pacific, 3 cm. **ID**: legs and claw arms green or reddish, banded with white lines

Squat shrimp (*Thor amboinensis*) Thoridae Indo-Pacific, 1,6 cm, on anemones, bubble corals. **ID**: orange with opaque white blue-bordered spots

SHRIMPS — BUMBLEBEE SHRIMPS, HARLEQUIN SHRIMPS

Cobourg's thorella (*Thorella cobourgi*) Hyppolytidae
West Pacific, 0,8 cm, on soft corals, rubble.
ID: thin translucent legs with red-dotted strips, whitish tail. IT

Banded tozeuma shrimp (*Tozeuma armatum*)
Hyppolytidae, Indo-Pacific, 5 cm, on black corals.
ID: elongated translucent body with irregular bands

Ocellated tozeuma shrimp (*Tozeuma lanceolatum*)
Hyppolytidae, West Pacific, 5 cm, on sponges, algae.
ID: acute rostrum, often with 2 ocelli on abdomen

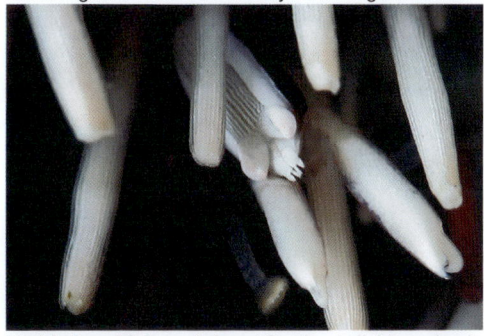

Urchin bumblebee shrimp (*Gnathophylloides mineri*)
Gnathophyllidae, Circumtropical, 1 cm, on sea urchins
ID: fine lines on the back, mimicking urchin's spines

Bumblebee shrimp (*Gnathophyllum americanum*)
Bumblebee shrimps (Gnathophyllidae)
Circumtropical, 2,5 cm

Harlequin shrimp (Hymenocera picta)
Hymenoceridae, 5 cm, Indo-Pacific. Feeds on sea stars. **ID**: White with big pink spots

Tiger shrimp (*Phyllognathia ceratophthalma*)
Hymenoceridae, Pacific, 2,5 cm. Feeds on ophiuras.
ID: translucent white, orange spots with blue dots

SHRIMPS — DRAGON, INSHORE, HINGE-BEAK SHRIMPS

Dragon shrimp (*Miropandalus hardingi*)
Pandalidae, Indo-Pacific, 3 cm, on black corals.
ID: forked rostrum

White-spotted inshore shrimp (*Neostylodactylus* sp.)
Stylodactylidae, West Pacific, 0,8 cm.
ID: translucent, with white areas to mimic rubble

Henderson's dancing shrimp (*Cinetorhynchus hendersoni*) Rhynchocinetidae, Indo-Pacific, 5 cm
ID: white with wavy red pattern. IT

Green-eyed dancing shrimp (*Cinetorhynchus reticulatus*) Rhynchocinetidae, Indo-Pacific, 5 cm
ID: red with round white spots. IT

Ocellated hinge-beak shrimp (*Rhynchocinetes conspiciocellus*) Rhynchocinetidae, West Pacific, 3,5 cm. ID: small white spots between red lines

Durban dancing shrimp (*Rhynchocinetes durbanensis*) Rhynchocinetidae, Indo-Pacific, 3 cm.
ID: unbroken white line between carapace and abdomen

Japanese night shrimp (*Hayashidonus japonicus*)
Night shrimps (Processidae) Indo-Pacific, 2,5 cm
ID: yellow, brown-bordered tail

Siboga night shrimp (*Nikoides sibogae*)
Processidae, Indo-Pacific, 2,5 cm
ID: translucent with red banding

SHRIMPS   PRAWNS (PENAEIDAE)

Night shrimp (orange-dotted bands) (Processidae sp.) Philippines, 3 cm
ID: translucent with slanted reddish bands. IT

Red-and-white night shrimp (*Processidae* sp.) Processidae, Philippines, 2,5 cm
ID: red claws, bent back

Long-arm prawn (*Heteropaenus longimanus*) Prawns (Penaeidae) West Pacific, 9 cm
ID: grey-brown spotting, good swimmer

Erythrean tiger shrimp (*Metapenaeopsis aegyptia*) Prawns (Penaeidae) Indo-Pacific, 8 cm
ID: red with white mottling

Humpback prawn (*Metapenaeopsis lamellata*) Prawns (Penaeidae) West Pacific, 10 cm
ID: prominent dentate rostrum

Red humpback prawn (*Metapenaeopsis* sp.) Prawns (Penaeidae) Indonesia, Philippines, 9 cm
ID: red with opaque white spots, short rostrum

Red-dotted humpback prawn (*Metapenaeopsis* sp.) Prawns (Penaeidae) Indonesia, Philippines, 5 cm
ID: prominent dentate rostrum, reddish with white spots

White-dotted humpback prawn (*Metapenaeopsis* sp.) Prawns (Penaeidae) West Pacific, 4 cm
ID: white saddle-like marks on abdomen. IT

SHRIMPS — PRAWNS, ROCK SHRIMPS, BOXER SHRIMPS

White-collar metapenaeopsis (*Metapenaeopsis* sp.)
Prawns (Penaeidae) Indonesia, Philippines, 6 cm
ID: short rostrum, white red-bordered carapace. **IT**

Western king prawn (*Penaeus latisulcatus*)
Prawns (Penaeidae) Indo-Pacific, 20 cm
ID: dark spots on the abdomen, blue tail-end

Two-spike sicyonia (*Sicyonia bispinosa*)
Rock shrimps (Sicyonidae) Indo-West Pacific, 2 cm
ID: prominent cylindrical rostrum

Japanese rock shrimp (*Sicyonia japonica*)
Rock shrimps (Sicyonidae) Indo-West Pacific, 4 cm.
ID: prominent dorsal teeth on carapace. **IT**

Red rock shrimp (*Sicyonia* sp.)
Rock shrimps (Sicyonidae) Indonesia, Philippines,
2,5 cm. **ID**: Reddish body, white tail

Crinoid boxer shrimp (*Odontozona* sp.)
Boxer shrimps (Stenopodidae) West Pacific, 2 cm.
ID: orange body, whitish claw arms

Banded boxer shrimp (*Stenopus hispidus*)
Boxer shrimps (Stenopodidae) Circumtropical, 6 cm
ID: red and white bands on the body and claw arms

Blue boxer shrimp (*Stenopus tenuirostris*)
Boxer shrimps (Stenopodidae) Indo-West Pacific,
2,5 cm. **ID**: Purple body, legs

LOBSTERS — SPINY LOBSTERS, SLIPPER LOBSTERS

DID YOU KNOW? Phyllosoma, a pelagic larvae stage of lobsters, utilizes jellyfish as food, transport, and shelter. **Picture: Liang Fu**

Fijian locust lobster (*Biarctus vitiensis*)
Slipper lobsters (Scyllaridae) Indo-Pacific, 5 cm
ID: white legs with dark bands. IT

Yellowbelly locust lobster (*Crenarctus bicuspidatus*)
Slipper lobsters (Scyllaridae) West Pacific, 7 cm
ID: yellowish belly, banded legs

Darkspot locust lobster (*Gibbularctus gibberosus*)
Slipper lobsters (Scyllaridae) West Pacific, 6 cm
ID: dark spot on the 1st abdominal segment. IT

Ornate spiny lobster (*Panulirus ornatus*)
Spiny lobsters (Palinuridae) Indo-Pacific, 70 cm
ID: black legs with white spots

Pronghorn spiny lobster (*Panulirus penicillatus*)
Spiny lobsters (Palinuridae) Indo-Pacific, 40 cm
ID: white-spotted abdomen, white lines down each leg

Painted spiny lobster (*Panulirus versicolor*)
Spiny lobsters (Palinuridae) Indo-Pacific, 90 cm
ID: white lines on abdominal segments and legs

MANTIS SHRIMPS (STOMATOPODA)

Red mantis shrimp (*Lysiosquillina lisa*)
Lysiosquillidae
Indo-West Pacific, 32 cm

Tiger mantis (*Lysiosquilla maculata*)
Lysiosquillidae, Indo-Pacific, 40 cm. **ID**: antennal scales with pepper-and-salt patch in the centre

White mantis shrimp (*Lysiosquilla* sp.)
Lysiosquillidae, Indonesia, Philippines, 15 cm. **ID**: translucent-bluish, white dots on carapace and claws

Golden mantis (*Lysiosquilloides mapia*)
Lysiosquillidae, West Pacific, 15 cm. **ID**: bright yellow, antennal scales with small golden spots

Keel tail mantis (*Odontodactylus cultrifer*)
Odontodactylidae, Indo-West Pacific, 13 cm. **ID**: blade-like red keel (larger in males than in females)

Pink-eared mantis (*Odontodactylus latirostris*)
Odontodactylidae, Indo-West Pacific, 8 cm. **ID**: bluish purple antennal scales in female, red in male

Peacock mantis shrimp (*Odontodactylus scyllarus*) Indo-Pacific, 18 cm. **ID**: dark spots on the carapace
DID YOU KNOW? Some species of mantis shrimps are monogamous, and can live together for more than 20 years. In some species, the female will lay two balls of eggs: one for male to tend and one for female

MANTIS SHRIMPS (STOMATOPODA) SEA SPIDERS (PYCNOGONIDA)

Bumptail mantis (*Haptosquilla tuberosa*)
Protosquillidae, Indo-West Pacific, 6 cm.
ID: big rounded granules on telson

Neptun's mantis shrimp (*Alima neptuni*)
Squillidae, Circumtropical, 7 cm.
ID: translucent with golden-dotted areas on carapace and abdomen

False mantis (*Pseudosquilla ciliata*) Pseudosquillidae, Circumtropical, 10 cm.
ID: Variable coloration: dark brown with pale central stripe, green, yellow, brown, black, cream. Checkerboard pattern on eyes

Hairy-legged sea spider (*Anoplodactylus* sp.)
Pycnogonida, West Pacific, 2,6 cm (including legs)
ID: opaque translucent body and legs

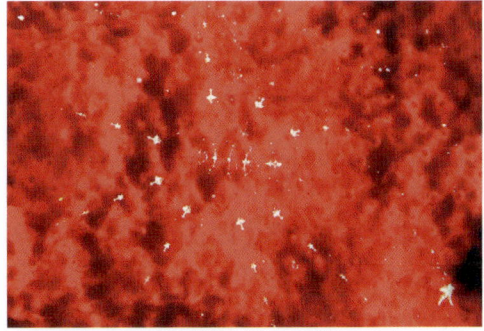

Transparent sea spider (*Parapallene* sp.)
Pycnogonida, West Pacific, 1,5 cm. **ID**: translucent, with whitish legs joints and head

85

CRABS-ANOMURA   SQUAT LOBSTERS (GALATHEIDAE)

Baba's crinoid squat lobster (*Allogalathea babai*)
Squat lobsters (Galatheidae) Indo-West Pacific, 2 cm. **ID**: longitudinal pale stripe down the carapace

Feather star squat lobster (*Allogalathea elegans*)
Squat lobsters (Galatheidae) Indo-Pacific, 2 cm
ID: longitudinal dark stripe down the carapace

Dark handcuff squad lobster (*Galathea cymo*)
Squat lobsters (Galatheidae) West Pacific, 1,5 cm
ID: dark bands behind yellow pincers

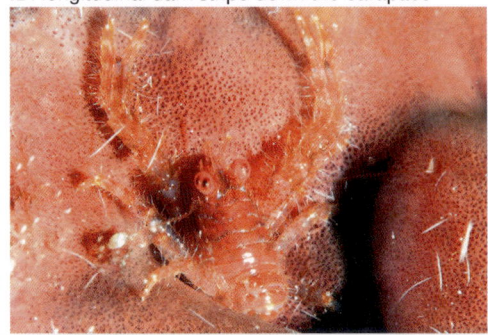

Red-face galathea (*Galathea* cf. *parvula*)
Squat lobsters (Galatheidae) Philippines, 1,5 cm
ID: translucent with orange spots and red eyes

Side-striped squat lobster (*Galathea* sp.)
Squat lobsters (Galatheidae) West Pacific, 1,7 cm
ID: white arrow down centre of sharp rostrum

Golden rings squat lobster (*Galathea* sp.)
Squat lobsters (Galatheidae) Philippines, 1,0 cm
ID: blue eyes, red pincers, golden banding on legs

Grooved squat lobster (*Galathea* sp.)
Squat lobsters (Galatheidae) West Pacific, 1,5 cm
ID: brown, long pointed roctrum, tips of pincers white

Yellow-dotted galathea (*Galathea* sp.)
Squat lobsters (Galatheidae) Philippines, 1,4 cm
ID: pale with red pattern, triangular red-dotted rostrum

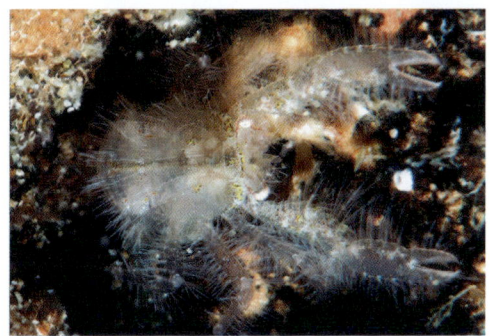

CRABS-ANOMURA SQUAT LOBSTERS, SPIDER SQUAT LOBSTERS

Hairy laurea (*Lauriea gardineri*)
Squat lobsters (Galatheidae) Indo-West Pacific, 1,5 cm
ID: scattered yellow black-dotted patches

Pink squat lobster (*Lauriea siagiani*)
Galatheidae, West Pacific, 1 cm. On barrel sponges, mimics purple brittle stars. **ID**: hairy, with purple outlines

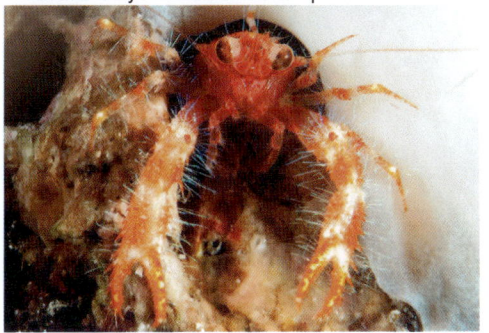

Olivar's squat lobster (*Munida olivarae*)
Squat lobsters (Munidae) West Pacific, 1,7 cm
ID: large grey eyes, short single rostrum spine

Blue-fingered squat lobster (*Sadayoshia tenuirostris*)
Munidae, Indo-West Pacific, 1,6 cm
ID: blue spot on the dactylus (movable finger)

White-handed squat lobster (*Sadayoshia* sp.)
Munidae, Philippines, 1 cm
ID: bright red-white pattern, red eyes. IT

Spider squat lobster (*Shirostylus sandyi*)
Chirostylidae, West Pacific, 1 cm (carapace)
ID: red or golden, spider-like legs with white dots

Gaimard's hermit crab (*Calcinus gaimardii*)
Diogenidae, Indo-West Pacific, 2 cm
ID: blue rings on the ends of orange eyestalks

Hidden hermit crab (*Calcinus latens*)
Diogenidae, Indo-Pacific, 1,5 cm
ID: white knobs on claws and legs, dark bands on legs

87

HERMIT CRABS LEFT-HANDED HERMIT CRABS (DIOGENIDAE)

Halloween hermit crab (*Ciliopagurus strigatus*)
Diogenidae, Indo-Pacific, 6 cm
ID: bright yellow and red banding, black-tipped fingers

Spotted black hermit crab (*Clibanarius cruentatus*)
Diogenidae, Indo-West Pacific, 1 cm
ID: beige eyestalks with white ends

White-dot-eyed clibanarius (*Clibanarius englaucus*)
Diogenidae, Indo-Pacific, 1,5 cm.
ID: beige eyestalks

White-dotted clibanarius (*Clibanarius snelliusi*)
Diogenidae, Indo-Pacific, 0,8 cm.
ID: appendages with white knobs, white eyestalks

Pale anemone hermit crab (*Dardanus deformis*)
Diogenidae, Indo-Pacific, 3 cm. **ID**: White eyestalks with dark band in the middle, green eyes

Reef hermit crab (*Dardanus lagopodes*)
Diogenidae, Indo-Pacific, 3 cm.
ID: legs, with dark bands on 1st segments

White-spotted hermit crab (*Dardanus megistos*)
Diogenidae, Indo-Pacific, 15 cm.
ID: deep red eyestalks

Anemone hermit crab (*Dardanus pedunculatus*)
Diogenidae, Indo-West Pacific, 5 cm.
ID: red and white striped eyestalks, green eyes

HERMIT CRABS RIGHT-HANDED HERMIT CRABS (PAGURIDAE)

Orange hermit crab (*Pseudopaguristes kuekenthali*)
Diogenidae, West Pacific, 1,5 cm
ID: orange with dark spots on the chelipeds

Coral hermit crab (*Paguritta harmsi*)
Paguridae, West Pacific, 1 cm.
ID: dark-green eyestalks with turquoise stripes

Distinct hermit crab (*Pagurixus dissimilis*)
Paguridae, Indo-West Pacific, 1,5 cm
ID: white appendages with brown rings

Hairy hermit crab (*Pagurus hirtimanus*)
Paguridae, Indo-Pacific, 3 cm.
ID: white dark-banded eyestalks, blue eyes

Blue-banded pagurus (*Pagurus* sp.) Paguridae
Philippines, 4 cm. **ID**: bluish eyes, white eyestalks crowned with dark stripe and blue outer part

Dark-dotted pagurus (*Pagurus* sp.) Paguridae
Philippines, 2 cm. **ID**: White appendages with small dark spots and grey stripes. IT

Tasseled hermit crab (*Pylopaguropsis fimbriata*)
Paguridae, West Pacific, 1,5 cm
ID: violet eyestalks, yellow eyes

Zebra hermit crab (*Pylopaguropsis zebra*)
Paguridae, West Pacific, 1,7 cm.
ID: maroon white-striped legs, pinkish right claw

CRABS-ANOMURA   PORCELAIN CRABS (PORCELLANIDAE)

Domed hermit crab (*Solitariopagurus trullirostris*) Paguridae, West Pacific, 1 cm. **ID**: white dotted translucent eyestalks with orange dash lines

Jumping hermit crab (*Spiropagurus spiriger*) Paguridae, West Pacific, 4 cm. Good jumper! **ID**: domed shells, blue eyes with dark spots

Barrel sponge porcelain crab (*Aliaporcellana spongicola*) Porcellanidae, Western Pacific, 1,5 cm, on barrel sponges

Soft coral porcelain crab (*Lissoporcellana nakasonei*) Porcellanidae, Indo-Pacific, 1 cm. **ID**: three slightly scalloped frontal teeth

Four-lobed porcelain crab (*Lissoporcellana quadrilobata*) Porcellanidae, IP, 1 cm. **ID**: three scalloped frontal teeth, prominent central tooth

Striped leg porcelain crab (*Lissoporcellana* sp.) Porcellanidae, West Pacific, 2 cm **ID**: reddish, pale banding on legs. IT

Dark-dotted porcelain crab (*Lissoporcellana* sp.) Porcellanidae, Philippines, PNG, 2 cm. **ID**: dark eyes, dark dots on the carapace and claw arms

Haig's porcelain crab (*Porcellanella haigae*) Porcellanidae, Indo-Pacific, 2 cm. **ID**: banded claws, pale carapace with brown stripes

CRABS -BRACHURA ROUND CRABS (XANTHIDAE)

Spotted porcelain crab (*Neopetrolisthes maculatus*) Porcellanidae
Red Sea, Indo-Pacific, 3 cm, on sea anemones. **ID**: White with small dark red dots, or larger red spots (right picture, **Oshimae porcelain crab**, *N. oshimai,* a synonym)

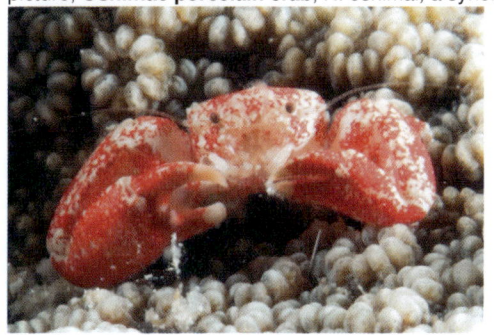

Red-white polyonyx (*Polyonyx biunguiculatus*)
Porcellanidae, West Pacific, 1,4 cm
ID: white with red-netted pattern

Pink-brown polyonyx (*Polyonyx cf. heok*)
Porcellanidae, Indonesia, Philippines, 1 cm
ID: pale brown, toothless rostrum. IT

Ruby serenius (*Serenius kuekenthali*)
Xanthidae, West Pacific, 3 cm
ID: bright red, lumpy claws with orange fingers

Red-eyed round crab (*Chlorodiella corallicola*)
Xanthidae, Indo-West Pacific, 3 cm. **ID**: pink or yellowish carapace with short chelae, banded legs

Hairy coral crab (*Cymo andreossyi*) Xanthidae
Indo-Pacific, 1,5 cm. **ID**: carapace and chelipeds with acute granules, white fingers

Amber etisus (*Etisus electra*) Xanthidae
Xanthidae, Indo-Pacific, 4 cm. **ID**: four prominent antero-lateral teeth, dark fingers

CRABS - BRACHURA   ROUND CRABS (XANTHIDAE)

Wrinkled round crab (*Hypocolpus pararugosus*)
Xanthidae, Indo-West Pacific, 5 cm. **ID**: carapace with white spots and tufts, red spots on claw arms

Orange medaeops (*Medaeops* cf. *edwardsi*)
Xanthidae, West Pacific, 2 cm.
ID: flat carapace, dark dotted claws, dark fingers

Tuberous xanthid crab (*Liomera monticulosa*)
Xanthidae, Indo-Pacific, 1,2 cm. **ID**: pale carapace with red spots on areas, covered with granules

Island land crab (*Neoliomera insularis*)
Xanthidae, Indo-West Pacific, 2,5 cm
ID: smooth red carapace, red legs with white banding

Anemone boxer-crab (*Lybia* sp.)
Xanthidae, West Pacific, 0,5 cm
ID: translucent legs with white spots

Mosaic boxer crab (*Lybia tessellata*) Xanthidae
Indo-Pacific, 2 cm
ID: brown and pink checkered pattern on carapace

Grooved demania (*Demania cultripes*) Xanthidae
West Pacific, 6 cm. **ID**: broad pink-brown rugose carapace with narrow clefts

Red-eyed actumnus (*Actumnus* sp.)
Hairy crabs (Pilumnidae), Philippines, 2 cm
ID: chelipeds with granules, red eyes with pale lines

CRABS - BRACHURA HAIRY CRABS, REEF CRABS, BOX CRABS

Pale hairy crab (*Caecopilumnus piroculatus*)
Pilumnidae, IWP, 2 cm. **ID**: convex carapace and chelipeds are covered by pale dense setae

Red-faced hairy crab (*Pilumnid* sp.)
Pilumnidae, West Pacific, 3 cm.
ID: round carapace with red areas under eyes. IT

Purple hairy crab (*Pilumnid* sp.) Pilumnidae
Philippines, 2 cm. **ID**: dark purple carapace and legs are matching with host sponge. IT

Zebra crab (*Zebrida adamsii*) Pilumnidae
Indo-Pacific, 2 cm, on sea urchins
ID: carapace with purplish-brown parallel bands

Variable coral crab (*Carpilius convexus*)
Reef crabs (Carpiliidae) Indo-Pacific, 12 cm
ID: smooth convex carapace with dark central spot

Spotted reef crab (*Carpilius maculatus*)
Reef crabs (Carpiliidae) Indo-Pacific, 20 cm
ID: smooth convex carapace with 11 red spots

Elongated box crab (*Calappa capellonis*)
Box crabs (Calappidae) West Pacific, 3 cm
ID: big rounded tubercles on carapace and claws

Rough box crab (*Calappa gallus*)
Box crabs (Calappidae) Indo-West Pacific, 10 cm.
ID: yellow legs, red or brown, lumpy carapace

CRABS - BRACHURA   SWIMMING CRABS (PORTUNIDAE)

Hepatic box crab (*Calappa hepatica*)
Box crabs (Calappidae) Indo-Pacific, 9 cm
ID: pale eyes with dark stripes on long stalks

Spotted box crab (*Calappa torulosa*) Box crabs (Calappidae) Indo-Pacific, 4 cm. **ID**: carapace with big round tubercles, often with dark areas

Ridged swimming crab (*Charybdis (Charybdis) natator*) Portunidae
Indonesia, Philippines, 20 cm

Oriental swimming crab (*Charybdis (Charybdis) orientalis*) Portunidae, Indo-Pacific, 4 cm.
ID: cheliped palm with 4 spines on upper border

Sea cucumber swimming crab (*Lissocarcinus orbicularis*) Portunidae, Indo-Pacific, 1,5 cm, on sea cucumbers
ID: carapace with spots, banded legs

Sea star swimming crab (*Lissocarcinus polybiodes*) Portunidae, Indo-Pacific, 2 cm, on sea stars.
ID: five antero-lateral teeth, banded legs

Philippines scissor swimming crab (*Lupocyclus philippinensis*) Portunidae, Indo-Pacific, 7 cm
ID: long thin claw arms

94

CRABS - BRACHURA   SWIMMING CRABS (PORTUNIDAE)

Rounded swimming crab (*Lupocyclus rotundatus*) Portunidae, Indo-Pacific, 3 cm **ID**: dorsal surface of carapace with distinct ridges

Silver swimming crab (*Portunus (Monomia) argentatus*) Portunidae, IP, 5 cm. **ID**: black and white spotted carapace with 9 antero-lateral teeth

Gladiator swimming crab (*Portunus (Monomia) gladiator*) Portunidae, IP, 12 cm. **ID**: broad carapace with distinct pattern of granular areas

Blue swimming crab (*Portunus* cf. *pelagicus*) Portunidae, Philippines, 15 cm. **ID**: carapace greenish-brown with dark areas on dorsal surface

Blue swimming crab (*Portunus pelagicus*) Indo-Pacific, 20 cm. **ID**: males brownish, carapace with pale spots, blue legs, females without blue tint

Blood-spotted swimming crab (*Portunus sanguinolentus*) Portunidae Red Sea, Indo-Pacific, 15 cm

Calfless swimming crab (*Portunus (Xiphonectes) tenuipes*) Portunidae, IP, 3 cm. **ID**: reddish or brown, front with 3 teeth, outer pair much broader

Black fingers swimming crab (*Thalamita aff. coeruleipes*) Portunidae, WP, 6 cm. **ID**: black-tipped antero-lateral teeth, orange marginal band

CRABS - BRACHURA   SWIMMING CRABS (PORTUNIDAE)

Shy swimming crab (*Thalamita danae*) Portunidae Indo-Pacific, 7 cm. ID: greenish-blue or blue, carapace with blue transverse ridges

Kagoshima's swimming crab (*Thalamita kagosimensis*) Portunidae, West Pacific, 4 cm. ID: round carapace with small white and bluish spots

Orange swimming crab (*Thalamita pelsarti*) Portunidae, IP, 10 cm. ID: brown-orange with blue spines and granules, legs blue or green-blue

Philippine thalamita (*Thalamita philippinensis*) Portunidae, West Pacific, 4 cm. ID: reddish with pale symmetrical areas, spiny claws

Four-lobed swimming crab (*Thalamita sima*) Portunidae, Indo-Pacific, 5 cm. ID: greenish-brown, carapace covered with tiny hairs

Spiny claw swimming crab (*Thalamita spinimana*) Portunidae, Indo-Pacific, 9 cm. ID: reddish, spiny chelipeds. Picture: **I.Klopunova**

Long-eyed swimming crab (*Thalamitoides quadridens*) Swimming crabs (Portunidae) Indo-Pacific, 5 cm

Tubastrea coral crab (*Quadrella boopsis*) Coral crabs (Trapezidae), IP, 2 cm, on Tubastrea corals ID: carapace with small tubercles and yellow eyes

96

CRABS - BRACHURA　　CORAL CRABS (TETRALIIDAE)

Black coral crab (*Quadrella maculosa*)
Coral crabs (Trapezidae) Indo-Pacific, 2 cm, on black corals

Two tooth guard crab (*Trapezia bidentata*)
Trapezidae, IP, 1,5 cm, on branching corals.
ID: pale with pink eyes or yellow, with black eyes

Pink trapeze crab (*Trapezia cymodoce*)
Trapezidae, Indo-Pacific, 2,5 cm. **ID**: row of red dots on the carapace, dark spot on the dactylus

Spotted-leg guard crab (*Trapezia guttata*)
Coral crabs (Trapezidae)
Indo-Pacific, 1,5 cm

Serene's black coral crab (*Quadrella serenei*)
Trapezidae, IP, 2 cm, on black corals, sponges
ID: yellow eyes, yellow banding on legs

Pale tetralia (*Tetralia glaberrimma*)
Tetraliidae Indo-Pacific, 1,5 cm.
ID: legs with red and dark dots

Black-line tetralia (*Tetralia nigrolineata*)
Tetraliidae, Indo-Pacific, 2 cm.
ID: frontal margin with thin blue lines

Smooth domecia (*Domecia glabra*)
Domeciidae, Indo-Pacific, 2 cm. **ID**: pale with white spines on the carapace, chelipeds and legs

CRABS - BRACHURA PURSE CRABS (LEUCOSIIDAE)

Slender arcania (*Arcania gracilis*) Leucosiidae Indo-Pacific, 2 cm. **ID**: carapace with five marginal spines and large red white-bordered ocellus

Spiny arcania (*Arcania novemspinosa*) Leucosiidae West Pacific, 2 cm. **ID**: granulous carapace with 11 large spinules

Grand myra (*Myra grandis*) Leucosiidae Indo-Pacific, 4 cm. **ID**: smooth, shiny carapace, pink with pale areas, pink legs with white banding

Blunt-pointed raylilia (*Raylilia coniculifera*) Leucosiidae, PNG, Indonesia, 1,2 cm. **ID**: merus with 2 red bands, 3 midlateral marginal spines

Pebble crab (*Leucosia anatum*) Leucosiidae Indo-West Pacific, 5 cm. **ID**: carapace with four dark red ring markings

Dark-spotted leucosia (*Leucosia craniolaris*) Leucosiidae, West Pacific, 4 cm. **ID**: olive-grey carapace with symmetrical brown spots

Specious purse crab (*Nucia speciosa*) Leucosiidae, Indo-Pacific, 4 cm. **ID**: white with small red granules, brown eyes

Red purse crab (*Urnalana* sp.) Leucosiidae, Philippines, 2,5 cm. **ID**: red carapace, white legs with red banding

CRABS - BRACHURA   ELBOW CRABS, IMITATOR CRABS

Two-horned gomeza (*Gomeza bicornis*) Corystidae Indo-Pacific, 3 cm. **ID**: orange, grey eyes with with 2 spines and 2 long feather-like antennae between

Rear-spined elbow crab (*Aulacolambrus hoplonotus*) Parthenopidae, Indo-Pacific, 10 cm **ID**: pale, chelipeds covered with spines

Spiny elbow crab (*Daldorfia rathbunae*) Parthenopidae, West Pacific, 4,3 cm. **ID**: pentagonal relatively smooth carapace

Long hands elbow crab (*Parthenope longimanus*) Parthenopidae, Indo-Pacific, 3,7 cm. **ID**: rhombic carapace, depressed slender legs

Big decorator crab (*Epialtidae* sp.) Epialtidae Philippines, 15 cm (carapace). **ID**: large pyriform carapace and massive legs, forked rostrum

Soft coral crab (*Hoplophrys oatesi*) Imitator crabs (Epialtidae) Indo-Pacific, 2 cm. On soft corals *Dendronephthya*

Arrowhead crab (*Huenia heraldica*) Imitator crabs (Epialtidae), Indo-Pacific, 1,5 cm **ID**: male - narrow, elongate triangular carapace, flat dorsal surface. Female carapace (right picture) has two pairs of lobate lateral outgrowths and is wider and quite different from the male carapace

CRABS - BRACHURA | DECORATOR CRABS

Conical spider crab (*Xenocarcinus conicus*)
Epialtidae, West Pacific, 1,5 cm, on black corals
ID: legs and carapace without tubercles

Depressed spider crab (*Xenocarcinus depressus*)
Epialtidae, West Pacific, 1,5 cm, on gorgonians
ID: legs and carapace with tubercles

Wire coral crab (*Xenocarcinus tuberculatus*)
Epialtidae, Indo-Pacific, 2 cm
On wire corals

Red algae spider crab (*Achaeus* sp.) Inachidae
Indo-Pacific, 1 cm. **ID**: triangular carapace, white eyes, legs and chelipeds decorated with algae. IT

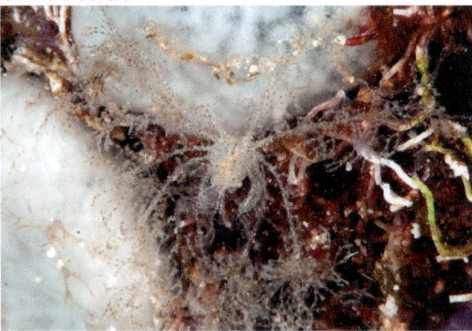

Curly spider-crab (*Chalaroachaeus curvipes*)
Inachidae, Indo-Pacific, 1 cm.
ID: translucent with dark dots, mimics bryozoan

Yellow spider-crab (*Oncinopus aranea*)
Inachidae, West Pacific, 2 cm. **ID**: puriform carapace, slender chelipeds, red white-dotted eyes

Orangutan crab (*Oncinopus* sp.)
Inachidae, West Pacific, 1,5 cm, on bubble corals
ID: covered by fine long hairs, red white-banded legs

Strawberry spider-crab (*Paratymolus* sp.)
Decorator crabs (Inachidae) West Pacific, 1 cm.
ID: round lumpy carapace with orange anterior area

CRABS - BRACHURA   STILT CRABS, GALL, SPONGE CRABS

Ornamental spider crab (*Schizophrys aspera*)
Majidae, Indo-Pacific, 6 cm. **ID**: pseudo-rostrum spines with one large accessory spine near the base

Pronghorn decorator crab (*Schizophrys dana*)
Majidae, Indo-Pacific, 6 cm. **ID**: pseudo-rostrum spines with two large accessory spines near the base

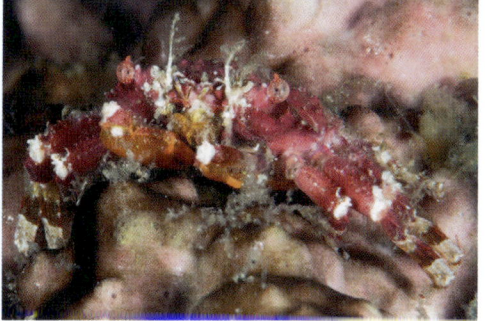

White-spotted pseudopalicus (*Pseudopalicus* sp.)
Stilt crabs (Palicidae) Indonesia, Philippines, 2 cm.
ID: red with white-banded legs, often yellowish claws

Greenish coral crab (*Pseudocryptochirus viridis*)
Gall crabs (Cryptochiridae) West Pacific, 0,6 cm
ID: yellow with turquoise orange-netted areas

Red sponge crab (*Cryptodromia coronata*)
Sponge crabs (Dromiidae) Pacific, 2 cm. **ID**: convex carapace with tiny tubercles, brown white-dotted eyes

Australian sponge crab (*Dromidiopsis australiensis*)
Dromiidae, West Pacific, 8 cm.
ID: red eyestalks with white triangular spot

Edward's sponge crab (*Dromidiopsis edwardsi*)
West Pacific, 18 cm. **ID**: convex carapace, brown eyes, whitish pincers, carry sponges

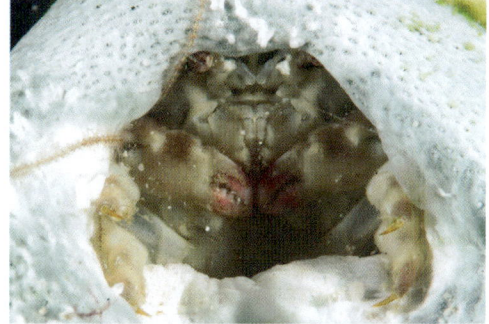

Plain-toothed sponge crab (*Lewindromia unidentata*)
Indo-Pacific, 2 cm. **ID**: small, with ascidians or sponges, smooth carapace, protruding rostrum

AMPHIPODS, COPEPODS

Golden metaprotella (*Metaprotella* cf. *sandalensis*) Caprellidae, Philippines, 1,5 cm. **ID**: red eyes, translucent body and legs with yellow intestines

White-eyed skeleton shrimp (*Protella* sp.) Caprellidae, Philippines, 1.5 cm. **ID**: translucent with red longitudinal lines

Tetragonal skeleton shrimp (*Quadrisegmentum* sp.) Caprellidae, Philippines, 1.1 cm. **ID**: smooth white body with tiny dark dots, pale eyes

Pipe amphipod (*Cerapus* sp.) Ischyroceridae, Philippines, 0.5 cm. Tube is made of minute sand grains and detritus held together with amphipod silk

Spiky-tail amphipod (*Iphimedia* sp.) Iphimediidae, Philippines, 5 mm. **ID**: bright yellow body and legs, antennas with white dots

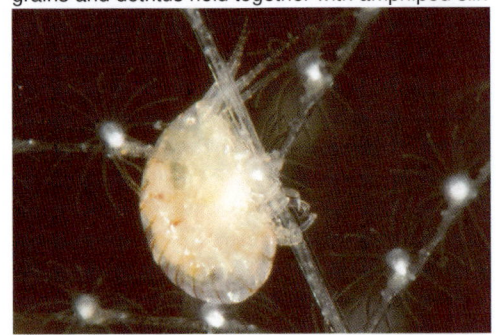

White hydroid amphipod (*Stenothoidae* sp.) Stenothoidae, Philippines, 2 mm, on hydroids **ID**: translucent with orange transversal lines

Social amphipod (*Ceradocus* sp.) Maeridae, Indonesia, Philippines, 1 cm, under stones **ID**: pinkish-orange body with pink eyes

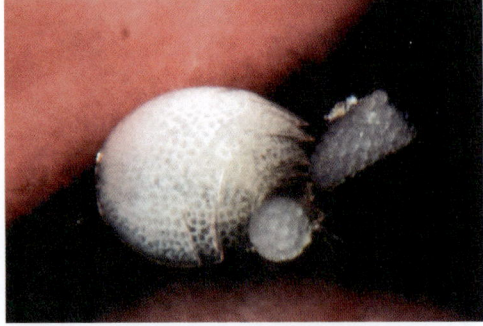

Minute copepod (*Harpacticoida* sp.) Harpacticoida West Pacific, 1 mm. Found on dead nudibranch. **ID**: tiny white copepod with eggs

ISOPODS, BARNACLES

Sponge isopod (*Santia* sp.)
Santiidae
Indo-Pacific, 3 mm, found on sponges

Pinkish sphaeromatid isopod (*Sphaeromatidae* sp.)
Sphaeromatidae, Philippines, 6 mm
ID: pinkish with dark-green dots and white areas

Striped isopod (*Sphaeromatidae* sp.) Sphaeromatidae
Philippines, 6 mm. **ID**: pinkish with wavy longitudinal lines crossed by transverse pale lines

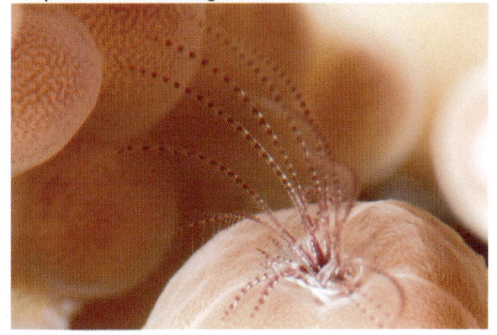

Acropora-dwelling barnacle (*Cantellius* sp.)
Pyrgomatidae, Philippines, 0,9 cm. **ID**: coral-dwelling barnacles, purple-white cirri (feeding legs)

Coral-dwelling barnacle (*Nobia grandis*)
Pyrgomatidae, Indo-West Pacific, 0,5 cm.
ID: picture shows biramous (forked) cirri

Married darwiniella (*Darwiniella conjugatum*)
Pyrgomatidae, Indo-West Pacific, 1 cm. **ID**: prefers *Cyphastrea* corals, translucent cirri

Dash-line barnacle (*Pyrgomatidae* sp.)
Pyrgomatidae, Philippines, 1 cm.
ID: cirri with black dashes and white dots

Orange-brimmed octolasmis (*Octolasmis* sp.)
Lepadidae, Philippines, 5 mm. **ID**: this small barnacles decided to live on the "face" of a huge crab

SEA SNAILS — COWRIES (CYPRAEIDAE)

Tiger cowry (*Cypraea tigris*)
Cowries (Cypraeidae) Indo-Pacific, 13,5 cm.
ID: mantle with white-tipped projections

Honey cowry (*Erosaria helvola*)
Cowries (Cypraeidae)
Indo-West Pacific, 4 cm

Labrolineata cowry (*Erosaria labrolineata*)
Cowries (Cypraeidae) Indo-Pacific, 4,3 cm.
ID: greenish siphon, pink tentacles

Millet cowry (*Erosaria miliaris*)
Cowries (Cypraeidae)
Indo-West Pacific Ocean, 4 cm

Fawn-coloured cowry (*Luria isabella*)
Cowries (Cypraeidae) Indo-Pacific, 5,4 cm.
ID: black matt semi-translucent mantle

Thick edged cowry (*Erronea caurica*)
Cowries (Cypraeidae) Indo-Pacific, 7 cm.
ID: greyish siphon and mantle, yellow tentacles

Carnelian cowry (*Lyncina carneola*)
Cowries (Cypraeidae)
Indo-Pacific, 9,5 cm

Lynx cowry (*Lyncina lynx*)
Cowries (Cypraeidae) Indo-Pacific, 8 cm. Mantle with white filaments, siphon and tentacles

SEA SNAILS   COWRIES (CYPRAEIDAE)

Ring top cowry (*Monetaria annulus*)
Cowries (Cypraeidae) Indo-Pacific, 5 cm
ID: shell with smoothly rounded sides and dorsum

Clandestine cowry (*Palmadusta clandestina*)
Cowries (Cypraeidae) Indo-Pacific, 2,7 cm.
ID: shell with fine orange banding, blue tentacles

Yellow cowry (*Palmadusta lutea*)
Cowries (Cypraeidae)
Indo-West Pacific, 2,7 cm

Zigzag cowry (*Palmadusta ziczac*)
Cowries (Cypraeidae)
Indo-Pacific, 2,6 cm

Fringed cowry (*Purpuradusta fimbriata*)
Cowries (Cypraeidae) I-P, 2,1 cm. **ID:** shell with purple spots at the anterior and posteriorends

Blotched cowry (*Purpuradusta gracilis*)
Cowries (Cypraeidae) Indo-Pacific, 3 cm
ID: brown mantle, yellow tentacles, grey siphon

Slug-like cowry (*Staphylaea limacina*)
Cowries (Cypraeidae) Indo-Pacific, 4 cm.
ID: long pinkish filaments

Tapering cowry (*Talostolida teres*)
Cowries (Cypraeidae)
Indo-Pacific, 4,4 cm

SEA SNAILS — FALSE COWRIES (OVULIDAE)

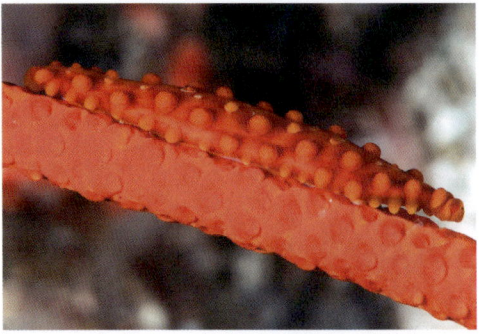

Lumpy spindle cowrie (*Aclyvolva lamyi*) False cowries (Ovulidae) West Pacific, 3 cm. Found on Dichotella octocorals. Feeds on host polyps.
ID: large domed tubercles

Umbilical egg shell (*Calpurnus verrucosus*) False cowries (Ovulidae) Indo-Pacific, 4 cm (shell), on Leather corals

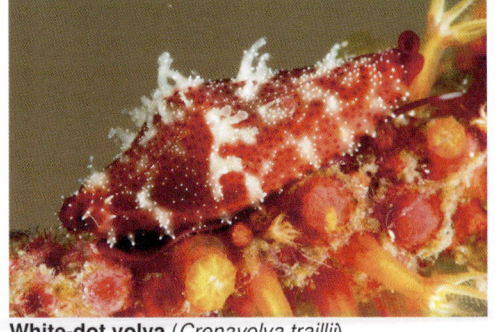

White-dot volva (*Crenavolva traillii*) False cowries (Ovulidae) West Pacific, 1,5 cm

Red mantle egg cowrie (*Cuspivolva* sp.) False cowries (Ovulidae) Philippines, 1 cm

Tiger egg cowry (*Cuspivolva tigris*) False cowries (Ovulidae) West Pacific, 1,7 cm (shell), feeds on Euplexaura sea fan

Margarita pearl ovulid (*Diminovula margarita*) False cowries (Ovulidae) West Pacific, 1 cm (shell)
ID: cylindrical papillae, rounded at the tip

White-tufted spindle cowry (*Hiatavolva rugosa*) False cowries (Ovulidae) Indo-West Pacific, 2,5 cm
ID: white tufts of papillae, mimicking white polyps

SEA SNAILS — FALSE COWRIES (OVULIDAE)

Canoe ovulid (*Naviculavolva deflexa*)
False cowries (Ovulidae) West Pacific, 3 cm (shell)
ID: white mantle with yellow margin

Pink-mouth egg shell (*Ovula costellata*)
False cowries (Ovulidae) Indo-Pacific, 4 cm (shell),
feeds on the soft coral *Sarcophyton* sp.

Common egg cowry (*Ovula ovum*)
False cowries (Ovulidae) Indo-Pacific, 12 cm, on
Leather corals. **ID**: black mantle with white papules

Graceful spindle cowry (*Phenacovolva gracilis*)
False cowries (Ovulidae) WP, 3,7 cm (shell). **ID**: long
tapering white papillae, pattern of red or brown lines

Gorgonian spindle cowry (*Phenacovolva rosea*)
False cowries (Ovulidae)
Indo-Pacific, 6,5 cm (shell)

Rosewater's volva (*Primovula rosewateri*)
False cowries (Ovulidae) Indo-West Pacific, 1,5 cm
(shell), on gorgonians

Fruit egg shell (*Prionovolva brevis*)
False cowries (Ovulidae) West Pacific, 2,5 cm
(shell), on Dendronephthya soft corals

Semistriated ovula (*Procalpurnus semistriatus*)
False cowries (Ovulidae) Indo-Pacific, 2,1 cm
(shell), on Leather corals

SEA SNAILS — MOON SNAILS (NATICIDAE)

Graceful moon snail (*Naticarius gracilis*)
Moon snails (Naticidae) West Pacific, 1,2 cm
ID: shell with fine transversal lines

Yolk moon snail (*Natica vitellus*)
Moon snails (Naticidae)
Indo-West Pacific, 5 cm

Butterfly moon snail (*Naticarius alapapilionis*)
Moon snails (Naticidae) West Pacific, 2,5 cm
ID: Red tentacles, foot with white bands

China moon snail (*Naticarius onca*)
Moon snails (Naticidae) Indo Pacific, 3,5 cm
ID: head shield with red blotches

Oriental moon snail (*Naticarius orientalis*)
Moon snails (Naticidae) Indo-West Pacific, 3,7 cm
ID: Yellowish shell, red white-banded foot

Nebulose moon snail (*Notocochlis cernica*)
Moon snails (Naticidae)
Indo-Pacific, 2,2 cm

Zebra moon snail (*Tanea undulata*)
Moon snails (Naticidae)
Indo-West Pacific, 3 cm

Lischke's tun (*Tonna lischkeana*)
Tonnidae (Tun shells) Indo-Pacific, 21 cm

108

SEA SNAILS — HELMETS, TRITONS, FROG, FIG SHELLS

Nodulose heavy bonnet (*Casmaria ponderosa nodulosa*) Helmets (Cassidae) West Pacific, 5,3 cm
ID: thin striped outer lip without denticles

Swollen casmaria (*Casmaria turgida*)
Helmetts (Cassidae)
West Pacific, 10 cm

Grey bonnet (*Phalium glaucum*)
Helmets (Cassidae) Indo-Pacific, 14,7 cm. **ID**: white or beige shell, white foot with wide brownish margin

Triton's trumpet (*Charonia tritonis*) Triton shells (Ranellidae), Indo-West Pacific, 50 cm. Feeds on the crown-of-thorns starfish, *Acanthaster planci*

Ruby triton (*Septa rubecula*)
Triton shells (Ranellidae)
Indo-West Pacific, 5,5 cm

Red-mouthed frog shell (*Tutufa rubeta*)
Bursidae (Frog shells)
Indo-Pacific, 25 cm

Paper fig shell (*Ficus ficus*)
Fig shells (Ficidae) Indo-West Pacific, 16,5 cm.
ID: pear-shaped thin shell with a long narrow aperture

Hoof scutus (*Scutus unguis*)
Keyhole limpets (Fissurellidae)
Indo-West Pacific, 4 cm

SEA SNAILS   CONCHES, EULIMIDS, RISSOIDS, VELUTINIDS

Minute conch (*Dolomena minima*)
Conches (Strombidae) West Pacific, 5 cm
ID: eyestalks and foot are green with white specks

Toothed conch (*Tridentarius dentatus*) Conches (Strombidae) Indo-Pacific, 6,5 cm. **ID**: three sharp projections from the outer lip near the anterior end

Spotted crinoid snail (*Annulobalcis maculatus*)
Eulimidae
West Pacific, 7 mm, parasitic on crinoids

Fire urchin snail (*Echineulima asthenosomae*)
Eulimidae, West Pacific, 1 cm, parasitic on fire urchins

Yellow spot rissoella (*Rissoella* sp.)
Rissoids (Rissoidae)
Philippines, 1 mm

Black velutinid (*Coriocella nigra*)
Velutinidae, Indo-Pacific, 10 cm
ID: brown eyes and tentacles with white tips

Golden sponge-mimic lamellaria (*Lamellaria* sp.)
Velutinids (Velutinidae)
Philippines, 1,5 mm

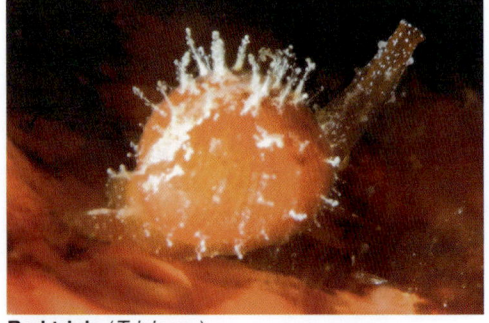

Red trivia (*Trivia* sp.)
Triviidae, West Pacific, 5 mm. Triviids are close to cowries, but only visually

SEA SNAILS — VOLUTES, HARPS, CONE SHELLS

Bat volute (*Cymbiola vespertilio*)
Volutes (Volutidae)
West Pacific, 16 cm

Huna's crithe (*Crithe huna*)
Cystiscidae (Cystiscid micromollusks)
Indo-Pacific, 0,4 cm

Articulate harp shell (*Harpa articularis*) Harp snails (Harpidae) I-W-P, 11 cm.
ID: massive foot with yellow dots

Checkered engina (*Engina spica*)
Pisaniidae, Indo-West Pacific, 1,3 cm
ID: brown-checkered pattern on the siphon and shell

Little frog cone (*Conus achatinus*)
Cone shells (Conidae)
Indo-Pacific, 10 cm

Geography cone (*Conus geographus*)
Cone shells (Conidae) Indo-Pacific, 16,6 cm.
Highly venomous, 30+ human fatalities registered

Fly-specked cone (*Conus stercusmuscarum*)
Cone shells (Conidae) West Pacific, 6,4 cm

Conch eucithara (*Eucithara stromboides*)
Mangeliidae, West Pacific, 2,5 cm

111

SEA SNAILS AUGERS, MITRE AND MUREX SNAILS, DOG WHELKS

Ribbed inquisitor (*Inquisitor sterrha*)
Pseudomelatomid snails (Pseudomelatomidae)
Indo-Pacific, 4 cm

Radish auger (*Duplicaria raphanula*)
Augers (Terebridae) Indo-Pacific, 8 cm

Succinct auger (*Terebra succincta*)
Augers (Terebridae)
Indo-Pacific, 6 cm

Flecked mitre (*Domiporta granatina*)
Mitre snails (Mitridae) Indo-Pacific, 7,4 cm
ID: Shell with intermittent brown spiral lines

Succory murex (*Hexaplex cichoreum*)
Murex snails (Muricidae)
West Pacific, 15 cm

Thin beak murex (*Murex ternispina*)
Murex snails (Muricidae)
Indonesia, Philippines, 13,8 cm

Papillose dog whelk (*Nassarius papillosus*)
Dog whelks (Nassariidae)
Indo-Pacific, 5 cm

Rosy phos (*Phos roseatus*)
Dog whelks (Nassariidae)
Indo-West Pacific, 4 cm

SEA SNAILS — WENTLETRAPS, TOP-SHELLS, TURBANS

Textile phos (*Phos textilis*)
Dog whelks (Nassariidae)
Indo-Pacific, 2 cm

Golden wentletrap (*Epitonium bileeanum*)
Wentletraps (Epitoniidae)
Indo-Pacific, 1,5 cm

Pyram wentletrap (*Epitonium pyramidale*)
Wentletraps (Epitoniidae)
Indo-Pacific, 5 cm

Secret clanculus (*Clanculus clanguloides*)
Top-shells (Trochidae)
West Pacific, 2 cm

Guam button top shell (*Ethalia guamensis*)
Top-shells (Trochidae)
Indonesia, Philippines, 2 cm

Tapestry turban (*Turbo petholatus*)
Turbans (Turbinidae)
Indo-West Pacific, 10 cm

Glistening abalone (*Haliotis glabra*) Abalone (Haliotidae) West Pacific, 7 cm. **ID**: flattened oval shell with 6-8 perforations and arrow-shaped spots

Perspective sundial (*Architectonica perspectiva*) Sundials (Architectonicidae) Indo-Pacific, 8,3 cm **ID**: white tentacles with black longitudinal stripes

MOLLUSCS — CHITONS, BIVALVES

Radial chiton (*Chiton densiliratus*)
Chitonidae, West Pacific, 4.8 cm.
ID tentative

Lamellar chiton (*Lucilina lamellosa*)
Chitonidae
Indo-West Pacific, 5 cm

Twilight chiton (*Lucilina* cf. *dilecta*)
Chitonidae
Philippines, 4 cm

Elongate chiton (*Schizochiton incisus*)
Schizochitonidae
Indo-West Pacific, 6 cm

Ventricose ark (*Arca ventricosa*)
Arcidae (Ark clams)
Indo-Pacific, 7 cm

White strawberry cockle (*Fragum fragum*)
Cardiidae (Cockle)
Indo-West Pacific, 4,5 cm

Mantis shrimp clam (*Scintilla* sp.)
Galeommatidae, Philippines, 1 cm.
Commensal with mantis shrimps

Cockscomb oyster (*Lopha cristagalli*)
True oysters (Ostreidae) Indo-West Pacific, 21 cm
ID: zig-zag pattern of the margins

MOLLUSCS — BIVALVES

Leaf oyster (*Dendostrea folium*)
True oysters (Ostreidae)
Indo-West Pacific, 14 cm

Variable oyster (*Spondylus varius*)
Thorny oysters (Spondylidae)
Indo-Pacific, 20 cm

Tiger venus clam (*Lioconcha tigrina*)
Venus clams (Veneridae) Indo-West Pacific, 2 cm, can jump by bending (left picture) and straightening (right picture) their foot

Lamellate venus clam (*Placamen lamellatum*)
Venus clams (Veneridae)
Indo-West Pacific, 4 cm

Large giant clam (*Tridacna maxima*)
Giant clams (Tridacninae) Indo-Pacific, 35 cm.
ID: partially embedded, weak vertical folds

Giant clam (*Tridacna gigas*)
Giant clams (Tridacninae) Indo Pacific, 135 cm.
The largest living bivalve mollusc

OPICTOBRANCHES   BUBBLE SNAILS, HEADSHIELD SLUGS

Brown-lined paperbuble (*Hydatina physis*)
Bubble snails (Aplustridae)
Circumtropical, 6 cm

Miniature melo (*Micromelo undatus*)
Bubble snails (Aplustridae)
Circumtropical, 3 cm

Red-lined bubble snail (*Bullina* sp.)
Bullinidae, West Pacific, 1,5 cm. Visually close to *Bullina nobilis*, Japanese species

Red-net diniatus (*Diniatys dubius*)
Haminoeid bubble snails (Haminoeidae)
West Pacific, 1,2 cm

Bottle bulla (*Bulla ampulla*)
Bullidae, West Pacific, 5 cm
ID: white spots and darker blotches

Striped atys (*Atys semistriata*)
Haminoeid bubble snails (Haminoeidae)
West Pacific, 8 mm

Long-tail haminoeid (*Haminoeid* sp.)
Haminoeid bubble snails (Haminoeidae)
West Pacific, 1 cm

Brown dots haminoeid (*Haminoeid* sp.)
Haminoeid bubble snails (Haminoeidae)
Philippines, 8 mm

OPICTOBRANCHES   HEADSHIELD SLUGS (CEPHALASPIDEA)

Red short-tail haminoeid (*Haminoeid* sp.)
Haminoeid bubble snails (Haminoeidae)
Philippines, 10 mm

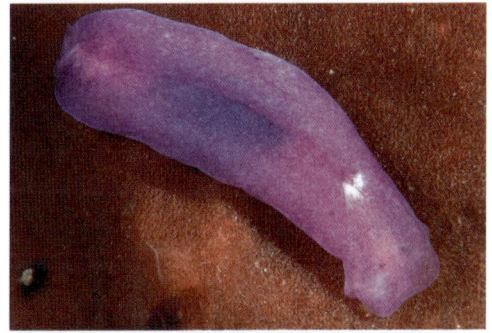

White-collar phanerophthalmus (*Phanerophthalmus albocollaris*) Haminoeid bubble snails (Haminoeidae)
West & Central Pacific, 1 cm

Fried eggs slug (*Colpodaspis thomsoni*)
Diaphanids (Diaphanidae)
Indo-Pacific, 3 mm

Lovely headshield slug (*Chelidonura amoena*)
Aglajidae
Indonesia, Philippines, 5,5 cm

Brilliant headsheild slug (*Chelidonura electra*)
Aglajidae, West Pacific, 8 cm.
ID: body with thin yellow margins

Yellow-spotted headshield slug (*Chelidonura fulvipunctata*) Aglajidae, West Pacific, 3 cm.
ID: white W-mark on the head

Swallowtail head shield slug (*Chelidonura hirundinina*)
Aglajidae, Circumtropical, 3 cm

Pale chelidonura (*Chelidonura pallida*)
Aglajidae
Indo-West Pacific, 5 cm

OPICTOBRANCHES   HEADSHIELD SLUGS (CEPHALASPIDEA)

Black chelidonura (*Chelidonura sandrana*)
Aglajidae
West Pacific, 2 cm

Blue velvet headshield slug (*Chelidonura varians*)
Aglajidae
West Pacific, 7 cm

Headband headsheild slug (*Mariaglaja inornata*)
Aglajidae, Western & Central Pacific, 5 cm.
ID: orange spots on the head

Frosted noalda (*Noalda* sp.)
Aglajidae
West Pacific, 7 mm

Guam odontoglaja (*Odontoglaja guamensis*)
Aglajidae
West Pacific, 1,5 cm

Raspberry philinopsis (*Philinopsis ctenophoraphaga*)
Aglajidae, Philippines, Indonesia, 2 cm

Yellow-dotted philinopsis (*Philinopsis falciphallus*)
Aglajidae
Philippines, Indonesia, 1 cm

Pilsbryi's philinopsis (*Philinopsis pilsbryi*)
Aglajidae
Red Sea, Indo-Pacific, 4 cm

OPICTOBRANCHES   HEADSHIELD SLUGS (CEPHALASPIDEA)

Red dot philinopsis (*Philinopsis* sp.)
Aglajidae
Philippines, Indonesia, Japan, 1,5 cm

Gardiner's headshield slug (*Tubulophilinopsis gardineri*)
Aglajidae
Indo-pacific, 3,5 cm

Reticulate philinopsis (*Tubulophilinopsis reticulata*)
Aglajidae, Indo-pacific, 3 cm. **ID**: blue markings on the anterior and posterior parts of the parapodia

White-dotted philinopsis (*Tubulophilinopsis* sp.)
Aglajidae, West Pacific, 3 cm.
ID: parapodia with yellow margin

Showy headshield slug (*Philinopsis speciosa*)
Aglajidae, Red Sea, Indo-Pacific, 3,5 cm. Coloration variable, feeds on opistobranches - Haminoeids, Sea hares

Blue-eared philinopsis (*Tubuphilinopsis* sp.)
Aglajidae, West Pacific, 1 cm.
ID: black with blue markings

Doublehorned batwing slug (*Gastropteron bicornutum*)
Gastropteridae, West Pacific, 1,5 cm

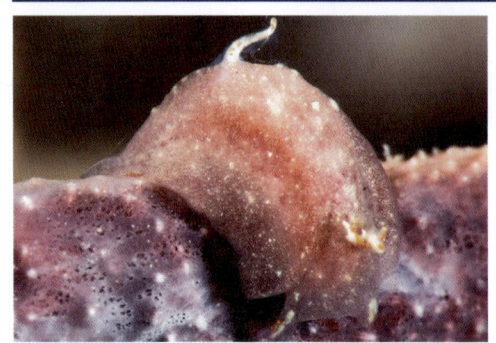

CEPHALASPIDEA   GASTROPTERIDAE

Orange-spotted batwing slug (*Sagaminopteron nigropunctatum*) Gastropteridae
Indo-West Pacific, 8 mm

White-margined batwing slug (*Sagaminopteron ornatum*) Gastropteridae
West Pacific, 1,5 cm

Psychedelic batwing slug (*Sagaminopteron psychedelicum*)
Gastropteridae, West Pacific, 2 cm

Dark-margined siphopteron (*Siphopteron brunneomarginatum*) Gastropteridae
West Pacific, 5 mm

Orange siphopteron (*Siphopteron flavolineatum*)
Gastropteridae
Philippines, 5 mm

Red-margin siphopteron (*Siphopteron* sp.)
Gastropteridae
Indo-Pacific, 5 mm

Milky hopper (*Siphopteron* sp.)
Gastropteridae, Philippines, 2 mm. **ID**: milky white, translucent grey tail with white spots. **IT. DBA**

Tiger siphonopteron (*Siphopteron tigrinum*)
Gastropteridae, West Pacific, 10 mm.
ID: purple marking, black-tipped siphon

SACOGLOSSA　　　　　　　OXYNOIDAE, CALIPHYLLIDAE

White-spotted lobiger (*Lobiger* sp.)
Oxynoidae, Indo-Pacific, 1,5 cm.
On *Caulerpa* green algae

Green lobiger (*Lobiger viridis*)
Oxynoidae, Indo-Pacific, 2 cm.
ID: blue lines on the shell

Tesselate oxynoe (*Oxynoe* sp.)
Oxynoidae, Western & Central Pacific, 1,5 cm.
On *Caulerpa filicoides* green algae.

Green oxynoe (*Oxynoe viridis*)
Oxynoidae, Indo-Pacific, 2 cm.
ID: round blue spots with yellow margin

Bourbon butterfly slug (*Cyerce bourbonica*)
Caliphyllidae, Indo-Pacific, 2 cm.
ID: small brown spots

Elegant cyerce (*Cyerce elegans*)
Caliphyllidae
Indo-Pacific, 5 cm

Dark butterfly slug (*Cyerce nigra*)
Caliphyllidae
West Pacific, 1,5 cm

Peacock butterfly slug (*Cyerce pavonina*)
Caliphyllidae, Indo-Pacific, 3 cm.
ID: cerata with transparent pustules

SACOGLOSSA — CALIPHYLLIDAE, COSTACIELLIDAE

White-capped cyerce (*Cyerce* sp.)
Caliphyllidae, Philippines, PNG, 2,5 cm.
ID: translucent cerata with white outlines

Reticulated butterfly slug (*Cyerce* sp.)
Caliphyllidae
West Pacific, 3 cm

Oriental polybranchia (*Polybranchia orientalis*)
Caliphyllidae, Indo-Pacific, 7 cm.
ID: cerata with visible digestive gland ducts

Kuro sapsucking slug (*Costasiella kuroshimae*)
Costasiellidae, Indo-Pacific, 10 mm. **ID:** sandglass-shaped orange patch above the rhinophores

Rabbit sapsucking slug (*Costasiella usagi*)
Costasiellidae
West Pacific, 1 cm

Yellow-capped stiliger (*Stiliger* sp.)
Costasiellidae
Western & Central Pacific, 8 mm

Dark-margined sapsucking slug (*Elysia marginata*)
Plakobranchidae, Indo-Pacific, 5 cm
ID: black spots, mantle edge orange and black

Halimeda sapsucking slug (*Elysia pusilla*)
Plakobranchidae, Indo-Pacific, 3,5 cm.
ID: white-tipped rhinophores

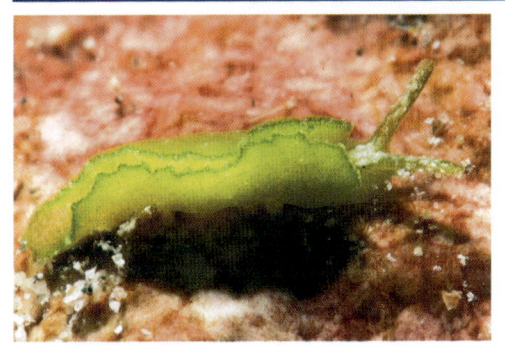

SACOGLOSSA   PLAKOBRANCHIDAE

White-dotted elysia (*Elysia* sp.)
Plakobranchidae
West Pacific, 7 mm

Ocellate plakobranchus (*Plakobranchus ocellatus*)
Plakobranchidae
Indo-Pacific, 6 cm

White-striped thuridilla (*Thuridilla albopustulosa*)
Plakobranchidae
West Pacific, 2,5 cm

Carlson's thuridilla (*Thuridilla carlsoni*)
Plakobranchidae
Western & Central Pacific, 3 cm

Yellow-spotted thuridilla (*Thuridilla flavomaculata*)
Plakobranchidae
West Pacific, 2 cm

Slender sapsucking slug (*Thuridilla gracilis*)
Plakobranchidae
Indo-Pacific, 2,5 cm

Hoff's thuridilla (*Thuridilla hoffae*)
Plakobranchidae
West Pacific, 2 cm

Wavy sapsucking slug (*Thuridilla undula*)
Plakobranchidae
Indo-Pacific, 1,5 cm

123

OPICTOBRANCHES   SEA HARES (ANASPIDEA)

Eyed sea hare (*Aplysia oculifera*)
Sea hares (Aplysiidae) Red Sea, Indo-Pacific, 8 cm
ID: small black rings

Ragged sea hare (*Bursatella leachii*)
Sea hares (Aplysiidae), Circumtropical, 15 cm.
ID: blue ocelli, long papillae

Freckled sea hare (*Aplysia parvula*)
Sea hares (Aplysiidae) Circumtropical, 7 cm. Feeds on red algae.
ID: parapodial flaps with dark margin

Wedge sea hare (*Dolabella auricularia*)
Sea hares (Aplysiidae)
Indo-Pacific & Eastern Pacific, 50 cm

White stripe seagrass slug (*Phyllaplysia* sp.)
Sea hares (Aplysiidae)
Philippines, 6 cm

Marten's sidegill slug (*Berthella martensi*)
Pleurobranchidae, Indo-Pacific & Eastern Pacific, 6 cm.
Feeds on sponges. Coloration variable

OPICTOBRANCHES　　　　　　　　　　SIDEGILL SLUGS

Starry berthella (*Berthella stellata*)
Pleurobranchidae. Circumtropical, 2,5 cm.
ID: white cruciform spot

Delicate berthellina (*Berthellina delicata*)
Pleurobranchidae
Indo-Pacific, 4 cm

Moon-headed sidegill slug (*Euselenops luniceps*)
Pleurobranchidae
Indo-Pacific, 7 cm

Brock's pleurobranch (*Pleurobranchaea brockii*)
Pleurobranchidae
Indo-Pacific, 12 cm

Forsskal's pleurobranch (*Pleurobranchus forskalii*)
Pleurobranchidae
Red Sea, Indo-Pacific, 30 cm

Grand pleurobranch (*Pleurobranchus grandis*)
Pleurobranchidae
Red Sea, Indo-Pacific, 20 cm

Peron's pleurobranch (*Pleurobranchus peronii*)
Pleurobranchidae, Indo-Pacific, 6 cm

Weber's pleurobranch (*Pleurobranchus weberi*)
Pleurobranchidae
West Pacific, 20 cm

NUDIBRANCHS — HEXABRANCHIDAE, POLYCERIDAE

Spanish dancer (*Hexabranchus sanguineus*) Hexabranchidae, Indo-Pacific, 60 cm
ID: Six separate gill branches that can not retract into a gill pouch
Right picture - juvenile

Ornate kalinga (*Kalinga ornata*)
Polyceridae, Indo-Pacific, 13 cm. Picture - sub-adult.

Shaggy kaloplocamus (*Kaloplocamus peludo*)
Polyceridae, West Pacific, 1,5 cm.
ID: covered with dark-brown dots

Pointed kaloplocamus (*Kaloplocamus* acutus)
Polyceridae
Indo-Pacific, 1,5 cm

Orange-dotted kaloplocamus (*Kaloplocamus* sp.)
Polyceridae
Philippines, 1 cm

Chamberlan's nembrotha (*Nembrotha chamberlaini*)
Polyceridae
Indo-Pacific, 6 cm

Crested nembrotha (*Nembrotha cristata*)
Polyceridae
Indo-Pacific, 12 cm

NUDIBRANCHS POLYCERIDAE

Variable neon slug (*Nembrotha kubaryana*)
Polyceridae
Indo-Pacific, 12 cm

Lined neon slug (*Nembrotha lineolata*)
Polyceridae, Indo-Pacific, 4 cm.
Picture: laying eggs, form mimics ascidian siphon

Livingstons nembrotha (*Nembrotha livingstonei*)
Polyceridae, West Pacific, 4 cm.
ID: white cross-like mark between the rhinophores

Milleri's nembrotha (*Nembrotha milleri*)
Polyceridae, West Pacific, 9 cm
Feeds on colonial ascidians

Mulliners nembrotha (*Nembrotha mullineri*) Polyceridae, West Pacific, 10 cm.
Coloration variable. Feeds on colonial tunicates
Left picture - juvenile

Ceylone plocamopherus (*Plocamopherus ceylonicus*) Polyceridae
Indo-Pacific, 2 cm

Wormwood plocamopherus (*Plocamopherus tilesii*) Polyceridae
West Pacific, 6 cm

NUDIBRANCHS   POLYCERIDAE

Dark-dash polycera (*Polycera* sp.)
Polyceridae, West Pacific, 1,2 cm.
ID: interrupted longitudinal dark lines

Yellow-frosted polycera (*Polycera* sp.)
Polyceridae
West Pacific, 9 mm

Gabrielas tambja (*Tambja gabrielae*)
Polyceridae
West Pacific, 6,5 cm

Gloomy tambja (*Tambja morosa*)
Polyceridae
Indo-Pacific, 10 cm

Painted thecacera (*Thecacera picta*)
Polyceridae
Indian ocean, Indo-West Pacific, 2 cm

Orange thecacera (*Thecacera* sp.)
Polyceridae
Philippines, Indonesia, Malaysia, 4 cm

Dark thecacera (*Thecacera* sp.)
Polyceridae
Indonesia, 4 cm

Purple pikachu (*Thecacera* sp.)
Polyceridae, Philippines, 4 cm.
ID: purplish with round orange spots

NUDIBRANCHS   GONIODORIDIDAE

Yellow-spotted thecacera (*Thecacera* sp.)
Polyceridae
Philippines, Taiwan, 9 mm

Brown-collar goniodoridella (*Goniodoridella* sp.)
Goniodorididae
Philippines, PNG, 4 mm

White goniodoris (*Goniodoris felis*)
Goniodorididae, West Pacific, 1 cm
ID: gill branches white with brown streaks

Brownspotted okenia (*Okenia brunneomaculata*)
Goniodorididae
Indonesia, Philippines, 1 cm

Japanese okenia (*Okenia japonica*)
Goniodorididae
West Pacific, 1,5 cm

Frilly okenia (*Okenia kendi*)
Goniodorididae, West Pacific, 2,5 cm
Named after the Tagalog word for candy

Red okenia *(Okenia kondoi)*
Goniodorididae, Philippines, 1,5 cm
ID: 4 pairs of dorsal appendages

Pink okenia (*Okenia* cf *liklik*)
Goniodorididae, Indo-Pacific, 6 mm
ID: lateral papillae without brown apices

129

NUDIBRANCHS — GONIODORIDIDAE, AEGIRETIDAE

Nakamoto okenia (*Okenia nakamotoensis*)
Goniodorididae, Indo-Pacific, 2 cm
ID: 5 pairs of dorsal appendages

Gilded trapania (*Trapania aurata*)
Goniodorididae
West Pacific, 1 cm

Paddle trapania (*Trapania palmula*)
Goniodorididae, West Pacific, 8 mm
ID: blue markings on the appendages

Jester trapania (*Trapania scurra*)
Goniodorididae
West Pacific, 1,5 cm

Soft coral trapania (*Trapania* sp.)
Goniodorididae, Philippines, 6-8 mm
Mimics soft coral polyps

Trapania filetto (*Trapania vitta*)
Goniodorididae
West Pacific, 1,2 cm

Banana nudibranch (*Notodoris minor*)
Aegiretidae
Indo-West Pacific, 10 cm

Grey norse god (*Aegires serenae*)
Aegiretidae, West Pacific, 9 cm
ID: yellow rhinophores

NUDIBRANCHS   AEGIRETIDAE, GYMNODORIDIDAE

Greenish aegires (*Aegires* sp.)
Aegiretidae
Philippines, 2 cm

Black-dottes aegires (*Aegires* sp.)
Aegiretidae, Philippines, 1,5 cm.
ID: tubercles with globular black tips

Pink aegires (*Aegires* sp.)
Aegiretidae
Indonesia, PNG, Philippines, 6 mm

Lavender aegires (*Aegires* sp.)
Aegiretidae
Indonesia, Philippines, 3 mm

Hairy norse god (*Aegires villosus*)
Aegiretidae
Indonesia, Philippines, Japan, 1,2 cm

Strawberry gymnodoris (*Gymnodoris aurita*)
Gymnodorididae
Indo-Pacific, 10 cm

Brownish gymnodoris (*Gymnodoris brunnea*)
Gymnodorididae, Philippines, 8 mm
ID: translucent, internal organs visible under gills

Ceylon gymnodoris (*Gymnodoris ceylonica*)
Gymnodorididae
Indo-Pacific, 12 cm

NUDIBRANCHS   GYMNODORIDIDAE

Unadorned gymnodoris (*Gymnodoris inornata*)
Gymnodorididae
Indo-Pacific, 6 cm

Orange-spotted gymnodoris (*Gymnodoris* sp.)
Gymnodorididae, WP, 1,5 cm.
ID: orange rhinophores.

Red bumpy gymnodoris (*Gymnodoris rubropapulosa*)
Gymnodorididae, Indo-Pacific, 6 cm

Orange-spotted gymnodoris (*Gymnodoris* sp.)
Gymnodorididae, Philippines, 1,5 cm
ID: gills forming a complete circle

Red-dash gymnodoris (*Gymnodoris* sp.)
Gymnodorididae, Philippines, 1,2 cm
ID: small orange and bigger milky-white spots

Translucent gymnodoris (*Gymnodoris* sp.)
Gymnodorididae, Philippines, 1,5 cm
ID: translucent, white rhinophores

Yellow-lined gymnodoris (*Gymnodoris* sp.)
Gymnodorididae
Philippines, Indonesia, 2 cm

Orange-lined gymnodoris (*Gymnodoris* sp.)
Gymnodorididae, Philippines, 3 cm
ID: close to previous one, generally bigger, red-orange

NUDIBRANCHS — DORIDIDAE, DISCODORIDIDAE

Ridged gymnodoris (*Gymnodoris subflava*)
Gymnodorididae
Philippines, Japan, 2 cm

Snow-flake gymnodoris (*Gymnodoris tuberculosa*)
Gymnodorididae
West Pacific, 5,5 cm

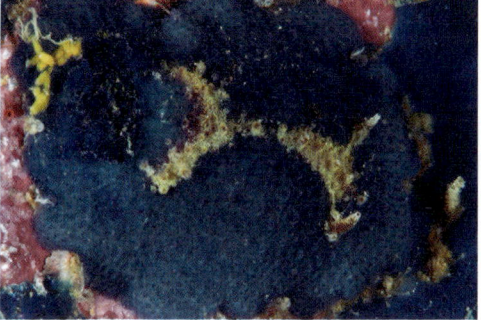

Blue-sponge doris (*Doris nucleosa*)
Dorididae, Indo-Pacific, 3 cm. Feeds on blue sponges, perfectly mimicking them

Blue doris (*Doris pecten*)
Dorididae, West Pacific, 2 cm
ID: blue all over, on blue sponges

Pink doris (*Doris* sp.)
Dorididae
Philippines, 2 cm

Lumpy asteronotus (*Asteronotus cespitious*)
Discodorididae
Indo-West Pacific, 40 cm

Liver-colored asteronotus (*Asteronotus hepaticus*)
Discodorididae, West Pacific, 55 cm.
Picture: subadult

White stripe atagema (*Atagema intecta*)
Discodorididae
Indo-Pacific, 8 cm

133

NUDIBRANCHS   DISCODORIDIDAE

Orange-brown dorid (*Caryophyllida dorid* sp.)
Discodorididae, Philippines, 2,5 cm
ID: brown with lighter stripe in the middle

Bohol discodoris (*Discodoris boholensis*)
Discodorididae
Indo-West Pacific, 7 cm

Batangas halgerda (*Halgerda batangas*)
Discodorididae
Indonesia, Malaysia, Thailand, 4 cm

Yellow halgerda (*Halgerda dalanghita*)
Discodorididae
Indo-West Pacific, 4 cm

Tesselated halgerda (*Halgerda tessellata*)
Discodorididae, West Pacific, 4 cm
ID: orange ridges on the mantle

Willeys halgerda (*Halgerda willeyi*)
Discodorididae, Indo-Pacific, 8 cm. **ID**: yellow-topped ridges, black and yellow lines between them

Mourning sea slug (*Jorunna funebris*)
Discodorididae
Red Sea, Indo-Pacific, 15 cm

Spotted jorunna (*Jorunna ramicola*)
Discodorididae
Indo-West Pacific, 2 cm

134

NUDIBRANCHS   DISCODORIDIDAE

Red-lined jorunna (*Jorunna rubescens*)
Discodorididae, Indo-Pacific, 20 cm.
ID: fine brown lines, yellow spots

Netted jorunna (*Jorunna* sp.)
Discodorididae, Philippines, 2 cm.
ID: pinkish with white network, mimicking sponges

Diagonal paradoris (*Paradoris* sp.)
Discodorididae
West Pacific, 3 cm

Dark-lined paradoris (*Paradoris* sp.)
Discodorididae
PNG, Philippines, 1,5 cm

Gray gill platydoris (*Platydoris cinerobranchiata*)
Discodorididae
West Pacific, 20 cm

Ocellate platydoris (*Platydoris ocellata*)
Discodorididae
West Pacific, 25 cm

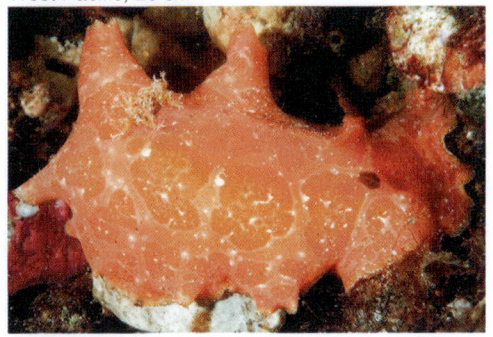

Grainy platydoris (*Platydoris* sp.)
Discodorididae
Philippines endemic, 6 cm

Leathery sclerodoris (*Sclerodoris* cf. *coriacea*)
Discodorididae, Philippines, 4 cm
Close to Indian Ocean species *S. coriacea*

NUDIBRANCHS — ACTINOCYCLIDAE, CHROMODORIDIDAE

Halgerda-lake taringa (*Taringa halgerda*)
Discodorididae
West Pacific, 5 cm

Bloody thordisa (*Thordisa sanguinea*)
Discodorididae
Philippines, Japan, 2 cm

Undecorated hallaxa (*Hallaxa indecora*)
Actinocyclidae
Indo-West Pacific, 1 cm

Avernis ardeadoris (*Ardeadoris averni*)
Chromodorididae
Indonesia, Philippines, PNG, 5,5 cm

Red spot ardeadoris (*Ardeadoris cruenta*)
Chromodorididae
West Pacific, 5,5 cm

Heron ardeadoris (*Ardeadoris egretta*)
Chromodorididae
Indonesia, Philippines, PNG, 10 cm

White ardeadoris (*Ardeadoris* cf *electra*)
Chromodorididae
Indo-Pacific, 4 cm

Ornamental cadinella (*Cadlinella ornatissima*)
Chromodorididae
Indo-Pacific, 3,5 cm

NUDIBRANCHS   CHROMODORIDIDAE

Brilliant ceratosoma (*Ceratosoma gracillimum*)
Chromodorididae, Indo-Pacific, 12 cm. **ID**: no mantle edge between the head and the lateral lobes

Red-lined ceratosoma (*Ceratosoma* sp.)
Chromodorididae
Indonesia, PNG, 2 cm

Many-lobed ceratosoma (*Ceratosoma tenue*)
Chromodorididae, Indo-Pacific, 12 cm.
ID: mantle with interrupted blue marginal line

Three-lobed ceratosoma (*Ceratosoma trilobatum*)
Chromodorididae, Red Sea, Indo-Pacific, 12 cm.
ID: mantle with purple marginal line

Anna's chromodoris (*Chromodoris annae*)
Chromodorididae, Indo-Pacific, 4 cm.
ID: black dots on blue areas

Purple-spotted chromodoris (*Chromodoris aspersa*)
Chromodorididae
Indo-Pacific, 2,6 cm

Coleman's chromodoris (*Chromodoris colemani*)
Chromodorididae
West Pacific, 2,5 cm

Elisabeth's chromodoris (*Chromodoris elisabethina*)
Chromodorididae, Indo-Pacific, 2,5 cm

NUDIBRANCHS   CHROMODORIDIDAE

Diana's chromodoris (*Chromodoris dianae*)
Chromodorididae, West Pacific, 6 cm. **ID**: 2 colour forms. First: white gills with an orange upper half (left)
Second: white gills with an orange edges (right picture)

Loch's chromodoris (*Chromodoris lochi*)
Chromodorididae
West Pacific, 3,5 cm

Magnificient chromodoris (*Chromodoris magnifica*)
Chromodorididae
West Pacific, 9 cm

Michaeli's chromodoris (*Chromodoris michaeli*)
Chromodorididae
West Pacific, 4,5 cm

Streaked chromodoris (*Chromodoris strigata*)
Chromodorididae
Red Sea, Indo-Pacific, 2 cm

Streaked chromodoris (*Chromodoris cf strigata*)
Chromodorididae, Indo-Pacific, 2 cm.
ID: yellow areas between dark lines

Willan's chromodoris (*Chromodoris willani*)
Chromodorididae, Indonesia, Philippines, 3,5 cm.
ID: white specks on the gills and rhinophores

NUDIBRANCHS   CHROMODORIDIDAE

Safron diversidorsis (*Diversidoris crocea*)
Chromodorididae
West Pacific, 2,5 cm

Large-eggs doriprismatica (*Doriprismatica balut*)
Chromodorididae, Philippines, Indonesia, 2 cm

Dark-margined doriprismatica (*Doriprismatica atromarginata*) Chromodorididae, Red Sea, Indo-Pacific, 10 cm

Spade-toothed doriprismatica (*Doriprismatica paladentata*) Chromodorididae, West Pacific, 2 cm
ID: bluish submarginal band

Yellow-margined doriprismatica (*Doriprismatica* sp.) Chromodorididae, Philippines, 4 cm. DBA. On the deep (40+ m) slopes

Margin glossodoris (*Glossodoris acosti*)
Chromodorididae, West Pacific, 4 cm. **ID**: marginal bands are wider, then in similar *G.* sp. cf. *cincta*

Coconut glossodoris (*Glossodoris buko*)
Chromodorididae, Western Pacific, 3 cm
Similar *G. pallida* appears to be Indian Ocean species

Brown margin glossodoris (*Glossodoris rufomarginata*)
Chromodorididae, Indo-Pacific, 4 cm

139

NUDIBRANCHS   CHROMODORIDIDAE

Coi's goniobranchus (*Goniobranchus coi*)
Chromodorididae
Western & Central Pacific, 5 cm

Collingwoods goniobranchus (*Goniobranchus collingwoodi*)
Chromodorididae, West Pacific, 4 cm

Creamy goniobranchus (*Goniobranchus fidelis*)
Chromodorididae
Indo-Pacific, 3 cm

Geometric goniobranchus (*Goniobranchus geometrica*) Chromodorididae, Indian Ocean, West Pacific, 3,5 cm

Bus stop hintuanensis (*Goniobranchus hintuanensis*)
Chromodorididae, Indo-Pacific, 1,6 cm

Kunie's goniobranchus (*Goniobranchus kuniei*)
Chromodorididae
Indo-Pacific, 3,5 cm

Leopard goniobranchus (*Goniobranchus leopardus*)
Chromodorididae
Indo-Pacific, 6 cm

Many-lined goniobranchus (*Goniobranchus lineolatus*) Chromodorididae, West Pacific, 3 cm.
ID: black with white lines and orange marginal band

NUDIBRANCHS   CHROMODORIDIDAE

Precious goniobranchus (*Goniobranchus preciosus*)
Chromodorididae, Indo-Pacific, 3 cm.
ID: marginal lines: white, dark ruby and orange

Spotted goniobranchus (*Goniobranchus reticulatus*)
Chromodorididae
West Pacific, 6 cm

Two-band goniobranchus (*Goniobranchus verrieri*)
Chromodorididae, Indo-Pacific, 1,7 cm
ID: 2 marginal lines: outer red and inner orange

Robe hem hypselodoris (*Hypselodoris apolegma*)
Chromodorididae, West Pacific Ocean, 10 cm
ID: reticulate white margin

Bullock's hypselodoris (*Hypselodoris bullocki*)
Chromodorididae, Western & Central Pacific, 4,5 cm
ID: white marginal line

Emmas hypselodoris (*Hypselodoris emma*)
Chromodorididae
Indo-Pacific, 4 cm

Fire hypselodorid (*Hypselodoris infucata*)
Chromodorididae, Red Sea, Indo-Pacific, 5 cm. **ID**: gills with red edge, mantle with blue and yellow spots

Spotted hypselodoris (*Hypselodoris maculosa*)
Chromodorididae, Indo-Pacific, 4 cm
ID: white with brown or purple nargin and white lines

NUDIBRANCHS CHROMODORIDIDAE

Red&black hypselodoris (*Hypselodoris purpureomaculosa*) Chromodorididae, West Pacific, 3,5 cm

Reids hypselodoris (*Hypselodoris reidi*) Chromodorididae, Indonesia, Philippines, 5 cm. **ID**: brown patch with white spots

Red-spotted hypselodoris (*Hypselodoris* sp.) Chromodorididae Indonesia, Philippines, 4 cm

White-margin hypselodoris (*Hypselodoris variobranchia*) Chromodorididae, Indonesia, Philippines, 5 cm.

Fish net hypselodoris (*Hypselodoris* jacula) Chromodorididae West Pacific, 8 cm

Yellow-spotted hypselodoris (*Hypselodoris* sp.) Chromodorididae, Indonesia, Philippines, 4,5 cm. **ID**: gills with purple ridges

Tryon's hypselodoris (*Hypselodoris tryoni*) Chromodorididae Western & Central Pacific, 7 cm

White's hypselodoris (*Hypselodoris whitei*) Chromodorididae Red Sea, Indo-Pacific, 3,5 cm

NUDIBRANCHS   CHROMODORIDIDAE

Small mexichromis (*Mexichromis pusilla*)
Chromodorididae
Indo-Pacific, 2 cm

Tree-lined mexichromis (*Mexichromis trilineata*)
Chromodorididae
West Pacific, 1,2 cm

Allens miamira (*Miamira alleni*)
Chromodorididae, Philippines, Indonesia, 8 cm.
Perfectly mimics soft coral with retracted polyps

Moloch miamira (*Miamira moloch*)
Chromodorididae, West Pacific, 11 cm.
Perfectly mimics sponge it feeds on

Netted miamira (*Miamira sinuata*)
Chromodorididae
West Pacific, 4,5 cm

Danielle's thorunna (*Thorunna daniellae*)
Chromodorididae, Indo-Pacific, 1,5 cm.
ID: diffused purple marginal band

Cryptic thorunna (*Thorunna furtiva*)
Chromodorididae
West Pacific, 2 cm

White-bordered thorunna (*Thorunna punicea*)
Chromodorididae
West Pacific, 1,2 cm

143

NUDIBRANCHS DENDRODORIDIDAE, PHYLLIDIIDAE

White-bordered noumea (*Verconia alboannulata*)
Chromodorididae
West Pacific, 2,5 cm

Blue-spot dendrodoris (*Dendrodoris denisoni*)
Dendrodorididae, Indo-Pacific, 6 cm
ID: brown areas with bright blue spots

Elongate dendrodoris (*Dendrodoris elongata*)
Dendrodorididae, Indo-Pacific, 6 cm.
ID: brown spots and white snowflake markings

Strawberry dendrodoris (*Dendrodoris guttata*)
Dendrodorididae
West Pacific, 6 cm

Black dendrodoris (*Dendrodoris nigra*)
Dendrodorididae, Indo-Pacific, 8 cm.
ID: adults black, only rhinophores with white tips

Black-ball ceratophyllidia (*Ceratophyllidia* sp.)
Phyllidiidae, West Pacific, 3,5 cm.
ID: globular papillae, white rhinophores

Baba's phyllidia (*Phyllidia babai*)
Phyllidiidae, West Pacific, 6,5 cm
ID: black rings with white tubercles

Sky blue phyllidia (*Phyllidia coelestis*)
Phyllidiidae, Indo-West Pacific, 6 cm.
ID: dark "Y" on the head

NUDIBRANCHS   PHYLLIDIIDAE

Elegant phyllidia (*Phyllidia elegans*)
Phyllidiidae, Indo-Pacific, 6 cm.
ID: white or yellow-capped tubercles

Exquisite phyllidia (*Phyllidia exquisita*)
Phyllidiidae, Indo-Pacific, 2,5 cm.
ID: yellow marginal line anteriorly

Eye spot sea slug (*Phyllidia ocellata*) Phyllidiidae, Indo-Pacific, 7 cm.
ID: coloration variable, yellow mantle with black circles or horseshoe-shaped markings

Painted phyllidia (*Phyllidia picta*)
Phyllidiidae
West Pacific, 4,5 cm

Polka dot phyllidia (*Phyllidia polkadotsa*)
Phyllidiidae
Indo-Pacific, 1,5 cm

Dark-lined phyllidia (*Phyllidia* sp.)
Phyllidiidae
Indonesia, Philippines, 6 cm

Waricose wart slag (*Phyllidia varicosa*)
Phyllidiidae, Indo-Pacific, 11,5 cm.
The most common *Phyllidia*

145

NUDIBRANCHS   PHYLLIDIIDAE

Willan's phyllidae (*Phyllidia willani*)
Phyllidiidae
West Pacific, 5 cm

Pimped phyllidiella (*Phyllidiella pustulosa*)
Phyllidiidae
Indo-Pacific, 7 cm

Cardinal phyllidiopsis (*Phyllidiopsis cardinalis*)
Phyllidiidae, Indo-Pacific, 6,5 cm.
ID: greenish rhinophores

Pipek's phyllidiopsis (*Phyllidiopsis pipeki*)
Phyllidiidae, West Pacific, 9 cm. **ID**: pink black-tipped rhinophores with a black line down posterior edge

Black-horned phyllidiopsis (*Phyllidiopsis sphingis*)
Phyllidiidae
West Pacific, 2 cm

Striped phyllidiopsis (*Phyllidiopsis striata*)
Phyllidiidae, Indo-West Pacific, 2 cm.
ID: 3 low white ridges

Mushroom coral reticulidia (*Reticulidia fungia*)
Phyllidiidae
Indo-Pacific, 4 cm

Reticulated halgerda (*Reticulidia halgerda*)
Phyllidiidae, Western & Central Pacific Oceans, 6,5 cm

NUDIBRANCHS   ARMINIDAE

Yellow-teeth armina (*Armina* cf. *comta*)
Arminidae, West Pacific, 4 cm
ID: oral veil with orange margin and "teeth"

Secret armina (*Armina occulta*)
Arminidae, West Pacific, 8 cm
ID: black foot with a blue edge

Blue veil armina (*Armina scotti*)
Arminidae, West Pacific, 6 cm
ID: foot and oral veil with yellow and blue marginal bands

Orange-margined armina (*Armina* sp.)
Arminidae, West Pacific, 10,5 cm
ID: foot and oral veil with orange margin

Orange-chin armina (*Armina* sp.)
Arminidae
Indonesia, Philippines, PNG, 4 cm

White dermatobranchus (*Dermatobranchus albus*)
Arminidae, Indo-West Pacific, 1 cm.
ID: mantle with distinct ridges

Blue-spotted dermatobranchus
(*Dermatobranchus caeruleomaculatus*)
Arminidae, Philippines, 2 cm

Blue-foot dermatobranchus (*Dermatobranchus cymatilis*)
Arminidae, Philippines, Japan, 2 cm.

147

NUDIBRANCHS DORIDOMORPHIDAE, PROCTONOTIDAE

Red-horned dermatobranchus (*Dermatobranchus funiculus*)
Arminidae, West Pacific, 1 cm

Ridged dermatobranchus (*Dermatobranchus gonatophorus*)
Arminidae, Indo-Pacific, 4 cm

Ornate dermatobranchus (*Dermatobranchus ornatus*)
Arminidae, West Pacific, 8 cm

Papillated dermatobranchus (*Dermatobranchus* cf. *pustulosus*)
Arminidae, Philippines, 7 cm

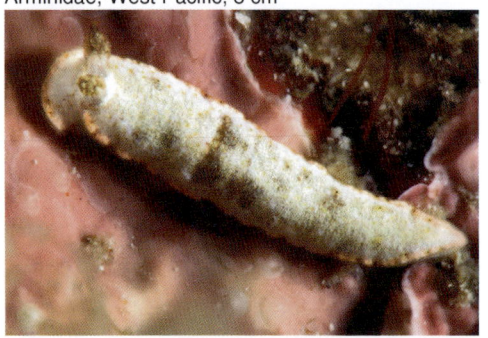

Rodman's dermatobranchus (*Dermatobranchus rodmani*)
Arminidae, Indo-Pacific, 1 cm

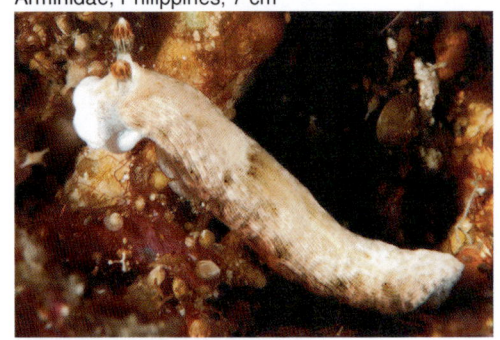

Orange-horned dermatobranchus (*Dermatobranchus* sp.)
Arminidae, Philippines, 1 cm

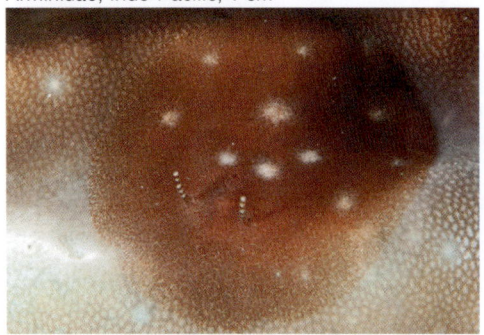

Blue-coral doridomorpha (*Doridomorpha gardineri*) Doridomorphidae, Indo-Pacific, 2 cm.
Found on broken blue corals, Heliopora caerulea

Savinkin's janolus (*Janolus savinkini*)
Proctonotidae
West Pacific, 7 cm

NUDIBRANCHS   BORNELLIDAE, LOMANOTIDAE, SCYLLAEIDAE

White spot janolus (*Janolus* sp.)
Proctonotidae, West Pacific, 2,5 cm.
ID: "frosted" white pigment

Eel bornella (*Bornella anguilla*)
Bornellidae, Indo-Pacific, 10 cm.
ID: black and orange-capped appendages

Hermann's bornella (*Bornella hermanni*)
Bornellidae, West Pacific, 5 cm.
ID: 3 pairs of appendages

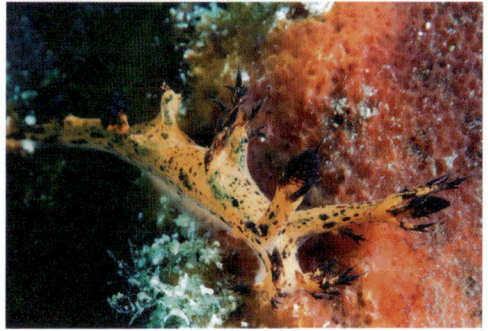

Regal dendronotus (*Dendronotus regius*)
Dendronotidae
Indonesia, Philippines, 3 cm

Netted lomanotus (*Lomanotus* sp.)
Lomanotidae, West & Central Pacific, 2 cm.
ID: brown with network of fine white lines

Flat back lomanotus (*Lomanotus* sp.)
Lomanotidae
West Pacific, 1 cm

Ward's notobryon (*Notobryon wardi*)
Scyllaeidae, West Pacific, 5 cm.
ID: iridescent blue tint on small papillae

Blue-dotted scyllaea (*Scyllaea fulva*)
Scyllaeidae, West Pacific, 4 cm.
Found on floating *Sargassum* algae

NUDIBRANCHS — THETYDIDAE, TRITONIIDAE

Ghost melibe (*Melibe colemani*)
Thetydidae, West Pacific, 5 cm.
ID: translucent, mimics Xenia soft corals

Branched melibe (*Melibe digitata*)
Thetydidae, West Pacific, 3,5 cm.
ID: brown, translucent, long rhinophores

Pink marianina (*Marianina rosea*)
Tritoniidae
Indo-West Pacific, 1,5 cm

Arborescent marionia (*Marionia arborescens*)
Tritoniidae, Indo-West Pacific, 6,5 cm.
Greenish gills, brownish reticulated notum

Divided marionia (*Marionia distincta*)
Tritoniidae, West Pacific, 6 cm.
ID: thin transverse brown lines and black spots

Greenish-yellow marionia (*Marionia elongoviridis*)
Tritoniidae
Philippines, 7 cm

Reticulate marionia (*Marionia* sp.)
Tritoniidae, Indonesia, Philippines, 7 cm.
ID: darker areas between the bases of the gills

Dark-stripe marionia (*Marionia* sp.)
Tritoniidae
West Pacific, 10 cm

NUDIBRANCHS   TRITONIIDAE

Strawberry marionia (*Marionia* sp.)
Tritoniidae
Indonesia, Philippines, 4 cm

Brown-gills marionia (*Marionia* sp.)
Tritoniidae, Oman, Philippines, 10 cm.
ID: network of thin white lines

Comb tritonia (*Tritonia* sp.)
Tritoniidae, Indo-West Pacific, perfectly mimics *Coelogorgia* host octocoral

Tile-patterned tritonia (*Tritonia* sp.)
Tritoniidae
West Pacific, 4 cm

Milky tritonia (*Tritonia* sp.)
Tritoniidae
Philippines endemic, 2,5 cm

Frozen tritonia (*Tritonia* sp.)
Tritoniidae
Philippines, 1,5 cm

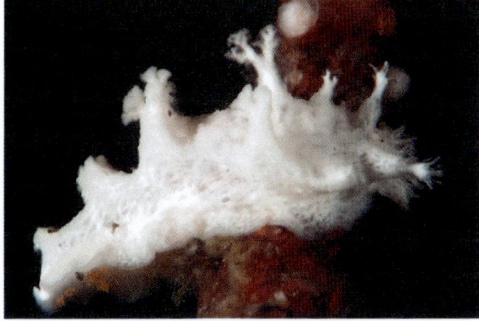

White tritoniopsis (*Tritoniopsis alba*)
Tritoniidae, Indo-Pacific, 3 cm.
ID: covered by small conical projections

Elegant tritonopsis (*Tritoniopsis elegans*)
Tritoniidae, Indo-Pacific, 6 cm.
ID: milky white spots on the dorsal surface

NUDIBRANCHS   DOTIDAE, EMBLETONIIDAE, FLABELLINIDAE

Blotched-face doto (*Doto* sp.)
Dotidae
Philippines, 2 cm

Racemose doto (*Doto racemosa*)
Dotidae
West Pacific, 2 cm

Yellow-tipped doto (*Doto* sp.)
Dotidae, Philippines, Indonesia, 2 cm. **ID**: brown-purplish rhinophores with yellow sheath margins

Grape doto (*Doto ussi*)
Dotidae, Indo-West Pacific, 3 cm.
ID: opaque white pigment

White kabeiro (*Kabeiro phasmida*)
Dotidae
West Pacific, 2 cm

Slender embletonia (*Embletonia gracilis*)
Embletoniidae
Indo-Pacific, 3 cm

Desirable coryphellina (*Coryphellina exoptata*)
Flabellinidae, Indo-Pacific, 2 cm.
ID: orange, papillate rhinophores

Purple aeolid (*Coryphellina* cf. *rubrolineata*)
Flabellinidae, Indo-Pacific, 2,5 cm
ID: papillate rhinophores, 3 red lines along the body

152

NUDIBRANCHS   SAMLIDAE, PICEINOTECIDAE, EUBRANCHIDAE

Bicolor samla (*Samla bicolor*)
Samlidae, Indo-Pacific, 1,5 cm.
ID: "frozen" white pigment

Double-ringed samla (*Samla bilas*)
Samlidae
Indo-Pacific, 2,5 cm

Orange-tipped samla (*Samla macassarana*)
Samlidae, Indo-West Pacific, 3 cm.
ID: cerata and rhinophores with diffuse orange tips

Netted samla (*Samla riwo*)
Samlidae, Indo-West Pacific, 2,5 cm.
ID: body with network of short dark lines

Yellow piseinotecus (*Piseinotecus* sp.)
Piseinotecidae
West Pacific, 1.2 cm

Papillate eubranchus (*Eubranchus mandapamensis*)
Eubranchidae, Indo-Pacific, 1,5 cm.
ID: cerata with yellow and blue bands and purple dots

Brown-spotted eubranchus (*Eubranchus* sp.)
Eubranchidae
Indonesia, 2,5 cm

Short-serata eubranchus (*Eubranchus* sp.)
Eubranchidae
West Pacific, 1,5 cm

NUDIBRANCHS TRINCHESIIDAE

Mournful phestilla (*Phestilla lugubris*)
Trinchesiidae, Indo-Pacific, 4 cm. On *Porites* corals.
ID: semi-translucent with opaque white spots

Pavona-eating phestilla (*Phestilla* sp.)
Trinchesiidae, Indo-West Pacific, 3 cm. Found on *Pavona* corals. **Picture**: male and female with eggs

Dark phestilia (*Phestilla melanobrachia*) Trinchesiidae, Indo-Pacific, 5 cm.
Feeds on dendrophyllid corals, often found under broken branches.
Cerata dark or orange

Goniopora-eating phestilla (*Phestilla* sp.)
Trinchesiidae, Indo-Pacific, 6 cm.
ID: fine transverse lines on the notum

Siboga trinchesia (*Trinchesia sibogae*)
Trinchesiidae
Indo-Pacific, 3,5 cm

Dark-ended trinchesia (*Trinchesia* sp.)
Trinchesiidae
Indo-Pacific, 5 mm

Yellow-masked trinchesia (*Trinchesia* sp.)
Trinchesiidae, West Pacific, 8 mm.
ID: cerata with yellow sub apical rings

NUDIBRANCHS TRINCHESIIDAE, FACELINIDAE

Long-horned trinchesia (*Trinchesia* sp.)
Trinchesiidae, Philippines, Indonesia, 1 cm
ID: cerata with red dots

Orange-horned trinchesia (*Trinchesia* sp.)
Trinchesiidae, West Pacific, 1 cm.
DBA, ID tentative

Yamasui's trinchesia (*Trinchesia yamasui*)
Trinchesiidae
Indonesia, Philippines, Japan, 15 mm

White-faced babakina (*Babakina indopacifica*)
Babakinidae
Indo-Pacific, 2 cm

Spider tergipes (*Tergiposacca longicerrata*) Fionidae, Indo-Pacific, 2 cm.
Usually found with white clusters of eggs.
ID: elongated cerata with digestive glands visible, coloration varies (yellow, orange, grey, burgundy

Indian caloria (*Caloria indica*)
Facelinidae
Indo-Pacific, 3 cm

White spot caloria (*Caloria* sp.)
Facelinidae
Philippines endemic, 2 cm

NUDIBRANCHS   FACELINIDAE

Tiger cratena (*Cratena simba*)
Facelinidae
Indian & West Pacific Oceans, 8 mm

Orange-specs cratena (*Cratena* sp.)
Facelinidae
West Pacific, 2 cm

White-dotted facelina (*Facelina* sp.)
Facelinidae
Philippines, 1 cm

Peach-colored facelina (*Facelina* sp.)
Facelinidae, West Pacific, 2 cm. **ID**: cerata and rhinophores pink-brown with white tips. **IT**

Rhodope facelina (*Facelina rhodopos*)
Facelinidae
West Pacific, 2 cm

Pink-topped facelina (*Facelina* sp.)
Facelinidae
West Pacific, 2 cm

Japanese favorinus (*Favorinus japonicus*)
Facelinidae, Indo-Pacific, 2 cm. **ID**: white mark with oblong spot behind the rhinophores

Wonderful favorinus (*Favorinus mirabilis*)
Facelinidae, Indo-Pacific, 1,2 cm
ID: yellow spot on the head

NUDIBRANCHS — FACELINIDAE

Red favorinus (*Favorinus* sp.)
Facelinidae, Philippines, 1,5 cm.
ID: oral tentacles with orange lines

Ringed favorinus (*Favorinus tsuruganus*)
Facelinidae, Indo-West Pacific, 3 cm.
ID: cerata with black apices

Diamond patch godiva (*Godiva* sp.)
Facelinidae
West Pacific, 6 cm

Orange moridilla (*Moridilla brockii*)
Facelinidae
Indo-West Pacific, 4 cm

Briarea phyllodesmium (*Phyllodesmium briareum*)
Facelinidae, Western Pacific, 2 cm.
Feeds on briareid soft corals

Cryptic phyllodesmium (*Phyllodesmium crypticum*) Facelinidae, Indo-West Pacific, 5 cm.
ID: nodulose cerata, dentate at the edges

Glassy phyllodesmium (*Phyllodesmium hyalinum*)
Facelinidae, Indo-West Pacific, 4,5 cm.
ID: cerata smooth, apically nodulose and curved

Jakobsens phyllodesmium (*Phyllodesmium jakobsenae*) Facelinidae, West Pacific, 3,5 cm.
Mimics its host prey *Xenia* soft corals

NUDIBRANCHS   FACELINIDAE

Kabira phyllodesmium (*Phyllodesmium kabiranum*)
Facelinidae
West Pacific, 4 cm

Koehlers phyllodesmium (*Phyllodesmium koehleri*)
Facelinidae, Western Pacific, 2 cm
ID: network of fine red lines on the notum

Solar powered phyllodesmium (*Phyllodesmium longicirrum*) Facelinidae, Western Pacific, 15 cm.
Feeds on leather corals

Undulate phyllodesmium (*Phyllodesmium undulatum*)
Facelinidae
West Pacific, 4 cm

Great phyllodesmium (*Phyllodesmium magnum*)
Facelinidae, Indo-Pacific, 18 cm.
ID: flattened curved cerata

White-midline phyllodesmium (*Phyllodesmium opalescens*)
Facelinidae, West Pacific, 2 cm

Fur-tree phyllodesmium (*Phyllodesmium pinnatum*)
Facelinidae
Philippines, 4 cm

Pink and purple phyllodesmium (*Phyllodesmium poindimiei*) Facelinidae, Indo-Pacific, 4 cm.
ID: coloration variable, white spots on the cerata

NUDIBRANCHS FACELINIDAE, AEOLIDIDAE

Rudman's phyllodesmium (*Phyllodesmium rudmani*)
Facelinidae, Indonesia, Philippines, 5 cm

Tuberculate phyllodesmium (*Phyllodesmium tuberculatum*) Facelinidae, Philippines, 3 cm.
ID: cerata with rounded tubercles

Blue dragon (*Pteraeolidia semperi*)
Facelinidae, Indo-Pacific, 10 cm.
ID: bright blue bands on the oral tentacles

Pink limenandra (*Limenandra barnosii*)
Aeolididae, Indo-West Pacific, 6 cm. Found at night near its prey - *Alicia* sea anemones

Moebii's baeolidia (*Baeolidia moebii*)
Aeolididae
Indo-Pacific, 7 cm

Affinitive cerberilla (*Cerberilla affinis*)
Aeolididae
Indo-Pacific, 9 cm

Annulate cerberilla (*Cerberilla annulata*)
Aeolididae
Indo-Pacific, 7 cm

White-dotted cerberilla (*Cerberilla* cf. *albopunctata*)
Aeolididae, Indonesia, Philippines, 2 cm
ID: grey cerata with yellow bands and white dots

CEPHALOPODS — CUTTLEFISHES, SQUIDS

Berry's bobtail squid (*Euprymna berryi*)
Cuttlefishes (Sepiidae)
West Pacific, 5 cm

White-eyed bobtail squid (*Sepiadarium kochi*)
Cuttlefishes (Sepiidae)
Indo-Pacific, 6 cm

Pfeffer's flamboyant cuttlefish (*Metasepia pfefferi*)
Cuttlefishes (Sepiidae), West Pacific, 8 cm. **Right picture**: adult. **Left picture**: 2 seconds after birth.
Research has shown the toxin in the muscle tissue to be as lethal as that of the blue-ringed octopus

Needle cuttlefish (*Sepia aculeata*)
Cuttlefishes (Sepiidae), Indo-West Pacific, 23 cm.
ID: triangular skin projections above fin

Dwarf cuttlefish (*Sepia bandensis*)
Cuttlefishes (Sepiidae), West Pacific, 5 cm.
ID: small blue spots on fin

Bigfin reef squid (*Sepioteuthis lessoniana*)
Squids (Teuthida)
Indo-Pacific, 40 cm

Algae octopus (*Abdopus aculeatus*)
Octopuses (Octopodidae), West Pacific, 25 cm.
Mimics gastropod shell with algae growth

CEPHALOPODS   OCTOPUSES

White-v octopus (*Abdopus* sp.)
Octopuses (Octopodidae), Indonesia, Philippines, 18 cm. **ID**: mantle end with white V-shaped spot

Coconut octopus (*Amphioctopus marginatus*)
Octopuses (Octopodidae), West Pacific, 30 cm.
ID: bluish sucker discs

Starry night octopus (*Callistoctopus luteus*)
Octopuses (Octopodidae), West Pacific, 80 cm.
ID: with white spots on legs and body

Greater blue-ringed octopus (*Hapalochlaena lunulata*)
Octopuses (Octopodidae), Indonesia, Philippines, Malaysia, 7 cm. Highly toxic! Don't touch!

Mimic octopus (*Thaumoctopus mimicus*)
Octopuses (Octopodidae), West Pacific, 30 cm.
ID: white outline under arms

Wunderpus octopus (*Wunderpus photogenicus*)
Octopuses (Octopodidae), West Pacific, 20 cm.
ID: reddish, without white outline under arms

Day octopus (*Octopus cyanea*)
Octopuses (Octopodidae), Indo-Pacific, 80 cm.
ID: large dark spot below the eye

Post-larvae, ready to settle octopus.
Philippines, 7-8 cm (head 1 cm)

ECHINODERMS   SEA STARS (ASTEROIDEA)

The crown-of-thorns starfish (*Acanthaster planci*) Acanthasteridae, Red Sea, Indo-Pacific, 60 cm.
Feeds on stony, coral polyps (Scleractinia), devastating huge reef areas.
Left picture - juvenile, 1 cm

Hefferman's starfish (*Fromia heffernani*)
Goniasteridae, West and Central Pacific, 10 cm.
ID: thin brown arms (normally 5)

Thousand-pores sea star (*Fromia milleporella*)
Goniasteridae, Indo-Pacific, 8 cm
ID: covered by tiny black pores or protrusions

Necklace sea star (*Fromia monilis*)
Goniasteridae, Red Sea, Indo-Pacific, 12 cm
ID: red central disc

Pacific sea star (*Fromia pacifica*)
Goniasteridae
Indo-Pacific, 7 cm

Unusual sea star (*Neoferdina insolita*)
Goniasteridae, West Pacific, 10 cm.
ID: upper arms outlined with red round spots

Ofrett's sea star (*Neoferdina offreti*)
Goniasteridae
Indo-Pacific, 7 cm

ECHINODERMS | SEA STARS (ASTEROIDEA)

Spiky gomophia (*Gomophia* sp.)
Ophidiasteridae
West Pacific, 14 cm

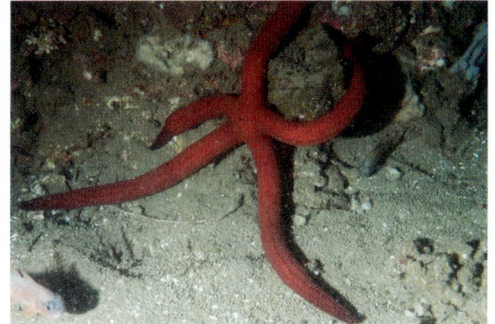

Velvet sea star (*Leiaster speciosus*)
Ophidiasteridae, Western and Central Pacific, 50 cm.
ID: rows of small dark papillae

Blue star (*Linckia laevigata*)
Ophidiasteridae
Indo-Pacific, 30 cm

Linckia sea star (*Linckia multifora*)
Ophidiasteridae, Red Sea, Indo-Pacific, 10 cm.
ID: scattered small red spots

New caledonian star (*Nardoa novaecaledoniae*)
Ophidiasteridae, Indonesia, Philippines, PNG, 12 cm

Original bothriaster (*Bothriaster primigenius*)
Cushion stars (Oreasteridae)
Indo-Pacific, 18 cm

Granulated sea star (*Choriaster granulatus*)
Cushion stars (Oreasteridae), Indo-Pacific, Red Sea, 27 cm

Cushion star (*Culcita novaeguineae*)
Cushion stars (Oreasteridae), Indo-West Pacific, 25 cm

ECHINODERMS — SEA STARS (ASTEROIDEA)

Chocolate chip sea star (*Protoreaster nodosus*)
Cushion stars (Oreasteridae), Indo-West Pacific, 30 cm

Astropecten sea star (*Astropecten* sp.)
Astropectinidae
West Pacific, 20 cm

Many-armed luidia (*Luidia avicularia*)
Luidiidae
West Pacific, 20 cm

Spotted luidia (*Luidia maculata*)
Luidiidae, West Pacific, 22 cm.
Feeds on brittle stars

Warty sea star (*Echinaster callosus*)
Echinasteridae, Indo-Pacific, 25 cm.
Feeds on algae and detritus

Luson sea star (*Echinaster luzonicus*)
Echinasteridae, Indonesia, Philippines, 15 cm.
ID: scattered dark pores

Slim euretaster (*Euretaster attenuatus*)
Pterasteridae
West Pacific, 15 cm

Striking sea star (*Euretaster insignis*)
Pterasteridae, West Pacific, 20 cm.
ID: net-like pattern

ECHINODERMS | BRITTLE STARS

Painted brittle star (*Ophiarthrum pictum*)
Ophiocomidae
Indo-Pacific, 20 cm

Variable brittle star (*Ophomastix variabilis*)
Ophiocomidae, Indo-Pacific, 9 cm.
ID: dark-brown to black with yellow markings

Six-armed brittle star (*Ophionereis* cf. *hexactis*)
Ophionereididae
West Pacific, 15 cm

Marten's brittle star (*Macrophiothrix martensi*)
Long-spined brittle stars (Ophiotrichidae) West Pacific, 20 cm

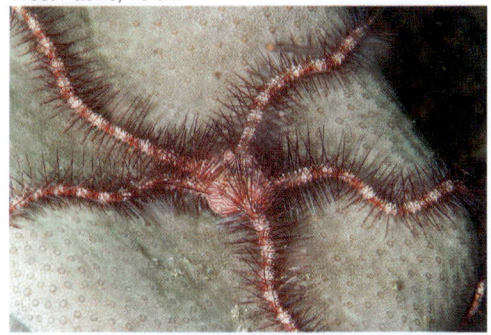

Purple brittle star (*Ophiothrix (Acanthophiothrix) purpurea*) Long-spined brittle star (Ophiotrichidae), Indo-Pacific, Red Sea, 15 cm

Deceptive brittle star (*Ophiothrix deceptor*)
Long-spined brittle stars (Ophiotrichidae)
Western & Central Pacific, 18 cm

Yellow-armed brittle star (*Ophiothrix* sp.)
Long-spined brittle stars (Ophiotrichidae)
West Pacific, 15 cm

Implicated brittle star (*Ophiarachna affinis*)
Ophiodermatidae
Red Sea, Indo-Pacific, 28 cm

ECHINODERMS — BRITTLE STARS, FEATHER STARS

Olive brittle star (*Ophiarachna incrassata*)
Ophiodermatidae, Indo-Pacific, 20 cm. Feeds on algae and small fishes, rapidly wrapping around them

Banded brittle star (*Ophiolepis superba*)
Ophiolepididae
Indonesia, Philippines, 12 cm

Spinose feather star (*Colobometra perspinosa*)
Colobometridae
Indo-Pacific, 20 cm

Noble feather star (*Comaster nobils*)
Colobometridae
Indo-Pacific, Red Sea, 40 cm

Variable bushy feather star (*Comaster schlegeli*)
Comasteridae
Indonesia, PNG, Philippines, 26 cm

Bennett's feather star (*Anneissia bennetti*)
Comasteridae
Indo-Pacific, 25 cm

Amazing clarkcomanthus (*Clarkcomanthus mirabilis*)
Comatulidae, West Pacific, 18 cm

Regal feather star (*Liparometra regalis*)
Mariametridae
West Pacific, 24 cm

ECHINODERMS   SEA URCHINS

Pencil sea urchin (*Phyllacanthus imperialis*)
Cidaridae, Red Sea, Indo-Pacific, 17 cm. Blunt spikes.
Feeds on sponges and algae

Vertical sea urchin (*Plococidaris verticillata*)
Cidaridae, West Pacific, 10 cm.
ID: dark star-like spot on the top

Raspy sea urchin (*Prionocidaris baculosa*)
Cidaridae, Indo-Pacific, 8 cm (test)
ID: collars of spines with bands of red spots

Rock-boring sea urchin (*Echinometra mathaei*)
Echinometridae
Indo-Pacific, Red Sea, 9 cm

Needle-spined urchin (*Echinostrephus aciculatus*)
Echinometridae
West Pacific, 18 cm

Globular sea urchin (*Mespilla globulus*)
Temnopleuridae
Indonesia, Philippines, 7,5 cm

Two-toned salmacis urchin (*Salmacis bicolor*)
Temnopleuridae, Indo-West Pacific, 15 cm.
ID: outer spines with red and white bands

Stained collector urchin (*Pseudoboletia maculata*)
Toxopneustidae, West Pacific, 11 cm.
ID: elongate reddish patches

ECHINODERMS SEA URCHINS (ECHINOIDEA)

Flower sea urchin (*Toxopneustes pileolus*)
Toxopneustidae, Indo-West Pacific, 20 cm.
Highly venomous, severe pain lasts about 1 hour

Collector urchin (*Tripneustes gratilla*)
Toxopneustidae
Indo-Pacific, Red Sea, 15 cm

Variable sea urchin (*Asthenosoma varium*)
Echinothuridae
Red Sea, Indo-Pacific, 22 cm

Radient sea urchin (*Astropyga radiata*)
Diadematidae
Indo-Pacific, Red Sea, 40 cm

Savigny's sea urchin (*Diadema savignyi*)
Diadematidae, Indo-Pacific, 23 cm.
ID: iridescent blue lines

Banded sea urchin (*Echinothrix calamaris*)
Diadematidae
Indo-Pacific, Red Sea, 30 cm

Reticulated sand dollar (*Clypeaster reticulatus*)
Sand dollars (Clypeasteridae)
West Pacific, 5 cm

Plain sand dollar (*Maretia planulata*)
Maretiidae
West Pacific, 9 cm

ECHINODERMS — SEA CUCUMBERS (HOLOTHUROIDEA)

Robust sea cucumber (*Colochirus robustus*)
Cucumariidae, West Pacific, 6 cm.
ID: feathery tentacles for catching zooplankton

Eyed sea cucumber (*Bohadschia argus*)
Holothuriidae
Indo-Pacific, Red Sea, 30 cm

Defensive sea cucumber (*Bohadschia vitiensis*)
Holothuriidae
Indo-Pacific, Red Sea, 24 cm

Yellow-spotted sea cucumber (*Bohadschia* sp.)
Holothuriidae, West Pacific, 30 cm. **ID**: dark brown, yellow papillae with black dot in the centre

Black sea cucumber (*Holothuria (Halodeima) atra*)
Holothuriidae, Indo-Pacific, Red Sea, 45 cm.
ID: entirely black, camouflaged by adhering sand

Edible sea cucumber (*Holothuria edulis*)
Holothuriidae
Indo-Pacific, Red Sea, 35 cm

Marbled sea cucumber (*Bohadschia marmorata*)
Holothuriidae
Indo-Pacific, 25 cm

Spiky holothuria (*Holothuria (Mertensiothuria) hilla*)
Holothuriidae, Indo-Pacific, 40 cm.
ID: brown with white conical papillae

ECHINODERMS — SEA CUCUMBERS (HOLOTHUROIDEA)

Elephant trunkfish (*Holothuria fuscopunctata*)
Holothuriidae, Indo-Pacific, 70 cm.
ID: dark spots and dark brown wrinkles

Kurt's holothuria (*Holothuria (Theelothuria) kurti*)
Holothuriidae
West Pacific, 24 cm

Translucent sea cucumber (*Labidodemas* sp.)
Holothuriidae
West Pacific, 15 cm

Graeffe's sea cucumber (*Pearsonothuria graeffei*)
Holothuriidae
Indo-Pacific, Red Sea, 40 cm

Reddish stichopus (*Stichopus* cf. *monotuberculatus*)
Holothuriidae, West Pacific, 35 cm

Amberfish sea cucumber (*Thelenota anax*)
Holothuriidae
Indonesia, Philippines, 90 cm

Candycane sea cucumber (*Thelenota rubralineata*)
Holothuriidae
West Pacific, 50 cm

Sticky snake sea cucumber (*Euapta godeffroyi*)
Synaptidae, Indo-Pacific, Red Sea, 40 cm
ID: fifteen feathery tentacles, longitudinal brown stripes

MARINE WORMS | ACOEL WORMS, FLATWORMS

Undescribed acoel worm (*Waminoa* sp.)
Acoel worms (Convolutidae) West Pacific, 1 cm.
ID: yellow dot at the base of caudal notch

White-brimmed acoel flatworm (*Waminoa* sp.)
Acoel worms (Convolutidae) Indonesia, Philippines, 1 cm.
ID: white submarginal band

Seaweed flatworm (*Gnesioceros sargassicola*)
Gnesiocerotidae
Pacific & Atlantic Ocean, 3 cm

Thick-bodied flatworm (*Callioplanidae* sp.)
Callioplanidae, West Pacific, 3 cm.
ID: network of fine red lines and small circles

Dark-middle cycloporus (*Cycloporus* sp.)
Polyclad flatworms (Euryleptidae)
West Pacific, 1.5 cm

Red-lined cycloporus (*Cycloporus* sp.)
Polyclad flatworms (Euryleptidae)
West Pacific, 2 cm

Gray flatworm (*Cycloporus* sp.)
Polyclad flatworms (Euryleptidae) West Pacific, 1 cm.
ID: small tentacles with orange margin

Red-neck eurylepta (*Eurylepta* sp.)
Polyclad flatworms (Euryleptidae) Philippines, 2 cm.
ID: orange patch around the base of the tentacles

FLATWORMS — EURYLEPTIDAE, PSEUDOCEROTIDAE

Brown-lined flatworm (*Prostheceraeus* sp.)
Polyclad flatworms (Euryleptidae) West Pacific, 3 cm
ID: white marginal line, orange-banded tentacles

Pijama prostheceraeus (*Prostheceraeus* sp.)
Polyclad flatworms (Euryleptidae)
West Pacific, 3 cm

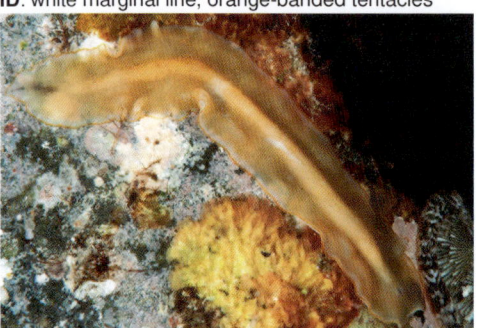

Delicate enchiridium (*Enchiridium* cf *delicatum*)
Prosthiostomid flat worms (Prosthiostomidae)
West Pacific, 3 cm

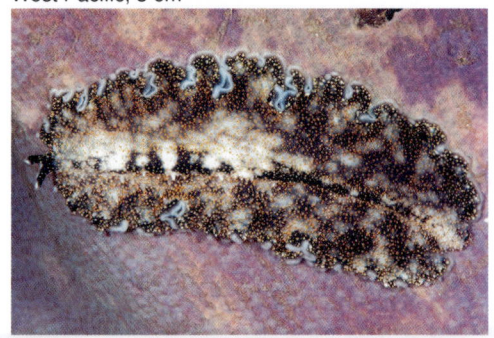

Brown-grey flatworm (*Thysanozoon* sp., poss. *Acanthozoon*) Pseudocerotidae, Indo-Pacific, 8 cm.
ID: small raised yellow-tipped papillae

Tunicate bulaceros (*Bulaceros* sp.)
Pseudocerotidae
West Pacific, 4 cm

Orange-netted bulaceros (*Bulaceros* sp.)
Pseudocerotidae
West Pacific, 4 cm

Orsak flatworm (*Maiazoon orsaki*)
Pseudocerotidae, Indo-Pacific, 4 cm.
ID: black marginal, diffuse orange submarginal line

Katoi's flatworm (*Phrikoceros katoi*)
Pseudocerotidae
Indo-Pacific, 1,4 cm

FLATWORMS   PSEUDOCEROTIDAE

Persian carpet flatworm (*Pseudobiceros bedfordi*)
Pseudocerotidae
Indo-West Pacific, 10 cm

Flower's flatworm (*Pseudobiceros flowersi*)
Pseudocerotidae, Indo-West Pacific, 4 cm.
ID: fine white marginal, diffuse black sub marginal line

Pleasing flatworm (*Pseudobiceros gratus*)
Pseudocerotidae
Indo-Pacific, 5 cm

One tree island flatworm (*Pseudobiceros hancockanus*) Pseudocerotidae, Indo-Pacific, 6 cm.
ID: 4 marginal lines - white, grey, orange, black

Cryptic flatworm (*Pseudobiceros kryptos*)
Pseudocerotidae
West Pacific, 4 cm

Gray flatworm (*Pseudobiceros murinus*)
Pseudocerotidae, West Pacific, 4 cm.
ID: small black dots. IT

White-margin pseudobiceros (*Pseudobiceros* sp.)
Pseudocerotidae
Indo-Pacific, 3 cm. IT

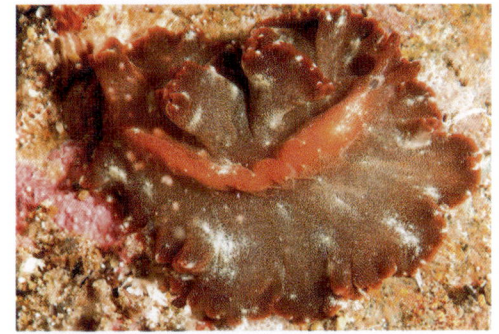

Reddish pseudobiceros (*Pseudobiceros* sp.)
Pseudocerotidae
West Pacific, 3 cm. IT

FLATWORMS　　　　　　　　　　　　　　　　　　PSEUDOCEROTIDAE

Racing stripe flatworm (*Pseudoceros bifurcus*)
Pseudocerotidae
Indonesia, Philippines, 6 cm

Sapphirine flatworm (*Pseudoceros caeruleocinctus*)
Pseudocerotidae, Indo-Pacific, 6 cm.
ID: 3 marginal bands, blue inner, black and thin white

Symmetrical flatworm (*Pseudoceros dimidiatus*)
Pseudocerotidae
Indo-West Pacific, 7 cm

Rusty flatworm (*Pseudoceros ferrugineus*)
Pseudocerotidae
West Pacific, Indian Ocean, 5 cm

Indian pseudoceros (*Pseudoceros indicus*)
Pseudocerotidae, Indo-Pacific, 4 cm.
ID: purple spots along margin

Speckled-edged pseudoceros (*Pseudoceros josei*)
Pseudocerotidae, West Pacific, 4 cm.
ID: dark broad marginal band with white dots. IT

Lindas flatworm (*Pseudoceros lindae*)
Pseudocerotidae
Indo-Pacific, 4,5 cm

Broadstriped flatworm (*Pseudoceros paralaticlavus*)
Pseudocerotidae
Western & Central Pacific, 5 cm

FLATWORMS PSEUDOCEROTIDAE

Red pseudoceros (*Pseudoceros rubroanus*)
Pseudocerotidae
West Pacific, 4 cm

Many-striped mimic worm (*Pseudoceros* sp.)
Pseudocerotidae, Philippines, 3 cm.
ID: broad orange marginal band

Chromodorid mimic flatworm (*Pseudoceros* sp.)
Pseudocerotidae, Indo-Pacific, 3 cm.
ID: white marginal band, 3 black stripes. **IT**

Chromodorid mimic flatworm (*Pseudoceros* sp.)
Pseudocerotidae, West Pacific, 2 cm.
Close to previous species, possibly colour form

Four-striped mimic worm (*Pseudoceros* sp.)
Pseudocerotidae, Philippines, 3 cm.
ID: broad orange marginal band, dark tentacles

Orange-banded flatworm (*Pseudoceros* sp.)
Pseudocerotidae, West Pacific, 3 cm.
ID: orange sub marginal band. **IT**

Brown flatworm (*Pseudoceros* sp.)
Pseudocerotidae
West Pacific, 4 cm

Wide-margined flatworm (*Pseudoceros* sp.)
Pseudocerotidae
West Pacific, 5 cm

RIBBON WORMS, PEANUT WORMS, FIRE WORMS

Hemprich's ribbon worm (*Baseodiscus hemprichi*)
Ribbon worms (Valenciniidae)
Indo-Pacific, 70 cm

Dark green ribbon worm (*Lineus fuscoviridis*)
Ribbon worms (Lineidae)
Indo-West Pacific, 25 cm

Peanut sipuncula worm (*Sipuncula* sp.)
Peanut worms (Sipuncula)
West Pacific, 4 cm

Amphora fire worm (*Chloeia amphora*)
Fire worms (Amphinomidae) Indo-Pacific, 10 cm.
ID: pattern of oval black spots with white outline

Amphora fire worm (*Chloeia* cf. *amphora*)
Fire worms (Amphinomidae) Philippines, 5 cm.
ID: pattern of oval black spots with white areas around

Conspicuous fire worm (*Chloeia conspicua*)
Fire worms (Amphinomidae) West Pacific, 6 cm.
ID: black rhombic pattern

Darklined fireworm (*Chloeia fusca*)
Fire worms (Amphinomidae) Indo-West Pacific, 8 cm.
ID: pattern of two central dark lines

Small fire worm (*Chloeia parva*)
Fire worms (Amphinomidae) Indo-Pacific, cm.
ID: caruncle with dark outline

SEGMENTED WORMS   ELONGATED, TUBE BUILDING WORMS

Orange fire worm (*Eurythoe complanata*)
Fire worms (Amphinomidae)
Circumtropical, 26 cm

White-blotched fire worm (*Notopygos* sp.)
Fire worms (Amphinomidae) Philippines, 2,5 cm.
ID: opaque white areas and white dots

Tile fire worm (*Pareurythoe* sp.)
Fire worms (Amphinomidae)
Philippines, 2 cm

Lined fire worm (*Pherecardia striata*)
Fire worms (Amphinomidae) Indo-Pacific, 20 cm.
ID: with dark-red stripes, forming longitudinal lines

Bobbit worm (*Eunice aphroditois*)
Large elongated worms (Eunicidae) Indo-Pacific, 1 m.
ID: dentate jaws, five banded tentacles

Yellow-dotted eunice (*Eunice* or *Leodice* sp.)
Large elongated worms (Eunicidae) West Pacific, 5 cm. **ID**: yellow spots on each segment

Oenonidae segmented worm (*Oenonidae* sp.)
Segmented worms (Oenonidae) West Pacific, 9 cm.
ID: flattened anterior part

Iridescent diopatra (*Diopatra* sp.)
Tube building worms (Onuphidae) West Pacific, 10 cm.
ID: dark dots on segments, 5 tentacles, red gills

MISCELLANEOUS SEGMENTED WORMS

Longibrachium tube worm (*Longibrachium* sp.)
Tube-building worms (Onuphidae)
West Pacific, 8 cm

Beautiful acoetid worm (*Acoetidae* sp.)
Acoetid worms (Acoetidae)
West Pacific, 10 cm

Sea cucumber scale worm (*Gastrolepidia clavigera*)
Scaled worms (Polynoidae) Indo-Pacific, 3 cm.
Found on echinoderms

Fifteen-scales scaleworm (*Harmothoe?* sp.)
Scaled worms (Polynoidae) Indonesia, 4 cm
ID: 15 pairs of elytra (scales) on the back. IT

Orange polynoid worm (*Polynoidae* sp.)
Scaled worms (Polynoidae)
West Pacific, 3 cm

Shaggy polynoid worm (*Polynoidae* sp.)
Scaled worms (Polynoidae)
West Pacific, 3 cm

Dark-banded hesione (*Hesione genetta*)
Hesionid worms (Hesionidae) West Pacific, 3 cm.
ID: red-brown bands

Black-and-white autolytinae worm (*Autolytinae* sp.)
Syllid worms (Syllidae)
Indonesia, Philippines, 5 cm

MISCELLANEOUS SEGMENTED WORMS

Grained scale worm (*Sphaerodoridae* sp.)
Sphaerodoridae
Philippines, 1 cm

Green notophyllum worm (*Notophyllum* sp.)
Phyllodocid worms (Phyllodocidae)
West Pacific, 3 cm

Yellow phyllodocid worm (*Phyllodocidae* sp.)
Phyllodocid worms (Phyllodocidae)
West Pacific, 3 cm

Delicate tube worm (*Filogranella elatensis*)
Calcareous tube worms (Serpulidae)
Indo-Pacific, 2,5 cm

Pale tube worm (*Hydroides* sp.)
Calcareous tube worms (Serpulidae)
Philippines, 2 cm

Magnificient tube worm (*Protula bispiralis*)
Calcareous tube worms (Serpulidae) Indo-Pacific, 4 cm.
ID: protruding tube, white and red colours

Christmas tree worm (*Spirobranchus* sp.)
Calcareous tube worms (Serpulidae) Indo-Pacific, 3 cm.
ID: coloration variable, operculum (tube cover) between spirals

MISCELLANEOUS SEGMENTED WORMS

Undescribed purplish duster worm (*Bispira* sp.)
Feather duster worms (Sabellidae)
West Pacific, 4 cm

Bluish duster worm (*Bispira* sp.)
Feather duster worms (Sabellidae)
Indo-West Pacific, 3 cm

Jelly-tube worm (*Myxicola* sp.)
Feather duster worms (Sabellidae) Indo-Pacific, 5 cm.
ID: horseshoe-shaped purple crown with white areas

Sanctijoseph's tube worm (*Sabellastarte sanctijosephi*) Feather duster worms (Sabellidae)
Indo-Pacific, 10 cm

Mason worm (*Lanice* sp.)
Spaghetti worms (Terebellidae) Indo-Pacific, 6 cm.
Tube composed of sand grains cemented with mucus

Greenish spaghetti worm (Terebellidae sp.)
Spaghetti worms (Terebellidae) West Pacific, 16 cm.
ID: greenish feeding tentacles

Southern horseshoe worm (*Phoronis australis*)
Horseshoe worms (Phoronida) Circumtropical, 3 cm.
Coloration variable, found near tube anemones

California horseshoe worm (*Phoronopsis californica*) Horseshoe worms (Phoronida)
Pacific Ocean, 4 cm

SEA SQUIRTS — CLAVELINIDAE, DIAZONIDAE

Distorted clavelina (*Clavelina detorta*)
Clavelinidae, Indo-Pacific, 3-6 cm (colony)
ID: yellow U-shaped intestine

Dark clavelina (*Clavelina fusca*)
Clavelinidae
West Pacific, 4 cm

Strong clavelina (*Clavelina robusta*)
Clavelinidae, Indo-Pacific, 3 cm.
ID: lemon bands around two siphons

Powdered tunicate (*Clavelina* sp.)
Clavelinidae, Philippines, 2 cm.
ID: golden specks on wall of zooids

Stalked green sea squirt (*Nephtheis fascicularis*)
Clavelinidae, Indo-West Pacific, 4 cm.
ID: fungi-like greenish clusters on stalks

White-spotted yellow tunicate (*Pycnoclavella diminuta*)
Clavelinidae
Indo-Pacific, 3-5 cm (colony)

Yellow pycnoclavella (*Pycnoclavella flava*)
Clavelinidae
Indo-Pacific, 1 cm

Nice ascidia (*Diazona formosa*)
Diazonidae tunicates (Diazonidae) West Pacific, 2 cm.
ID: six white spots around the cloacal siphon

SEA SQUIRTS   DIDEMNIDAE, POLYCITORIDAE, HOLOZOIDAE

Green barrel sea squirt (*Didemnum molle*)
Didemnidae, Indo-Pacific, 10 cm.
ID: numerous tiny oral siphons, one large cloacal siphon

Robust sea squirt (*Atriolum robustum*)
Didemnidae, Indo-Pacific, 3 cm.
ID: oral siphons relatively large

Blue-white tunicate (*Didemnid* sp.)
Didemnidae
Philippines, 11 cm

Green tunicate (*Lissoclinum patella*)
Didemnidae, Indo-Pacific, 8-20 cm (colony)
ID: opaque white ridges, green depressions

Vase tunicate (*Eudistoma laysani*)
Polycitoridae
Philippines, Indonesia, 4 cm

Yellow-edged stalk tunicate (*Sycozoa* sp.)
Holozoidae, West Pacific, 6 cm.
ID: two triangular yellow spots between siphons

Green-ringed ascidian (*Sigillina signifera*)
Holozoidae
Indo-Pacific, 5-20 cm (colony)

Pyramid sea squirt (*Aplidium breviventer*)
Polyclinidae
Indo-Pacific, 3 cm

SEA SQUIRTS ASCIDIIDAE, STYELIDAE

Ornate ascidia (*Ascidia ornata*)
Ascidiidae, Indo-Pacific, 5 cm
ID: yellow basket-like line pattern inside

Orange botrylloides (*Botrylloides* cf. *leachii*)
Styelidae, West Pacific, 10-14 cm (colony)
Small zooids are arranged in chains or networks

Dark colonial tunicate (*Botrylloides* cf. *leachii*)
Styelidae
Philippines, 10 cm

Dark botrylloides (*Botrylloides nigrum*)
Styelidae
Indo-Pacific, 30 cm (colony)

Red-chain tunicate (*Botrylloides* sp.)
Styelidae
Philippines, 10 cm (colony)

White-orange colonial tunicate (*Botrylloides* sp.)
Styelidae
Philippines, 7-10 cm (colony)

Golden star ascidian (*Botryllus* sp.)
Styelidae
Philippines, 9 cm (colony)

Black-and-white botryllus (*Botryllus sp.*)
Styelidae
West Pacific, 10-15 cm (colony)

SEA SQUIRTS   STYELIDAE

Greenish tunicate (*Botryllus* sp.)
Styelidae
Philippines, 5-12 cm (colony)

Gray tunicate (*Botryllus* sp.)
Styelidae
Philippines, 6-8 cm (colony)

Golden - stitched tunicate (*Botryllus* sp.)
Styelidae
Philippines, 7 cm (colony)

Brown-netted ascidian (*Botryllus* sp.)
Styelidae
Philippines, 14 cm (colony)

White-orange colonial ascidian (*Botryllus* sp.)
Styelidae
Philippines, 12 cm (colony)

Bricky eusynstyela (*Eusynstyela latericius*)
Styelidae
Indo-Pacific, 20 cm (colony)

Gold-mouth sea squirt (*Polycarpa aurata*)
Styelidae, Indo-Pacific, 15 cm.
ID: bright yellow inside

Apricot tunicate (*Polycarpa contecta* cf)
Styelidae
Indo-Pacific, 9 cm (colony)

MISCELLANEOUS SEA SPONGES (PORIFERA)

Spreading spiky sponge (*Callyspongia* sp.)
Callyspongiidae
Indo-West Pacific, 90 cm

Red spiky tube sponge (*Callyspongia* sp.)
Callyspongiidae, West Pacific, 28 cm.
ID: pink sponge with tube and encrusting forms. IT

Finger sponge (*Haliclona (Gellius) amboinensis*)
Chalinidae
West Pacific, 40 cm. IT

Crater sponge (*Hemimycale* sp.)
Hymedesmiidae, Atlantic Ocean, Mediterranean Sea,
Indo-West Pacific, 20 cm. IT

Octopus arm sponge (*Phoriospongia* sp.)
Chondropsidae
Philippines, 20 cm. IT

Granulated tube sponge (*Liosina granularis*)
Dictyonellidae, West Pacific, 45 cm (colony).
ID: tube sponge, smooth surface with irregular pits

Perforated veined sponge (*Clathria* sp.)
Microcioninae, Indo-West Pacific.
ID: flat, encrusting, with perforated depressions

Toothed barrel sponge (*Gelliodes fibulatus*)
Petrosiidae
Indo-Pacific, 20 cm

MISCELLANEOUS SEA SPONGES (PORIFERA)

Big barrel sponge (*Xestospongia testudinaria*)
Petrosiidae, West Pacific, 1 m.
ID: pink ridged sponge, found in current-prone areas

Tuberous ball sponge (*Diacarnus bellae*)
Podospongiidae
Indo-Pacific, 10 cm

Thick-veined sponge (*Aplysilla* sp.)
Darwinellidae
Indo-Pacific, 25-30 cm (colony)

Gray net sponge (*Chelonaplysilla* sp.)
Darwinellidae
Philippines, 10 cm

Ridged sponge (*Carteriospongia* cf. *contorta*)
Thorectidae, Indo-West Pacific, 25 cm.
ID: tubular appertures and radiating ridges

Shiny tubes sponge (*Leucandra palaoensis*)
Grantiidae, Indo-Pacific, 2 cm.
ID: coloration variable, network of shining tubes

Inclined calcite sponge (*Heteropia minor*)
Heteropiidae
Indo-West Pacific, 8 cm

Yellow lobe-sponge (*Leucetta* cf. *chagonensis*)
Leucettidae
Indo-Pacific, 20 cm

186

BRYOZOA BUGULIDAE, LANCEOPORIDAE, PHILODOPORIDAE

Brown bryozoan (*Bugula neritina*)
Bugulidae, Circumtropical, 8 cm (colony)
ID: purple zooids, 20-24 white, translucent tentacles

Maple leaf bryozoan (*Lanceopora* sp.)
Lanceoporidae
Indo-West Pacific, 4-7 cm

Hairy bryozoan (*Celleporaria sibogae*)
Lepraliellidae, Indo-Pacific, 12-15 cm (colony) with hydroid *Zanclea divergens*

Plate bryozoan (*Disporella* sp.)
Lichenoporidae, Philippines, 2-3 cm.
ID: round colonies with tubular zooids

Purple bryozoan (*Iodictyum sanguineum*)
Phidoloporidae, Indo-Pacific, 4-5 cm colony.
ID: purple network

White bryozoan (*Reteporellina* sp.)
Phidoloporidae
West Pacific, 6-9 cm. IT

Decorated bryozoan (*Triphyllozoon inornatum*)
Phidoloporidae
Indo-Pacific, 9-12 cm

Lacy napkin bryozoan (*Steginoporella* sp.)
Steginoporellidae
West Pacific, 10 -12 cm

187

CNIDARIANS   OCTOCORALS

Carnation soft coral (*Dendronephthya* sp.)
Soft corals (Nephtheidae) West Pacific, 30 cm.
ID: burgundy and yellowish polyps. IT

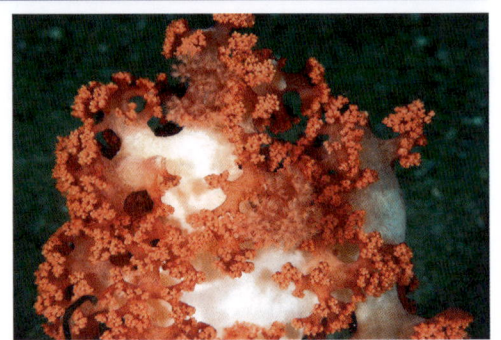

Orange soft coral (*Scleronephthya* sp.)
Soft corals (Nephtheidae) 15 cm.
ID: milky-white robust stalks, orange polyps. IT

Orange-mouthed soft coral (*Scleronephthya* sp.)
Soft corals (Nephtheidae)
West Pacific, 15-20 cm

Cherry blossom coral (*Siphonogorgia* cf. *godeffroyi*) Nidaliiae, Indo-Pacific, 60 cm.
ID: red smooth branches, yellow polyps

Blue cespitularia (*Cespitularia coerulea*)
Xeniidae, 15-20 cm.
ID: blue tentacles

Rusty heteroxenia (*Heteroxenia* sp.)
Xeniidae, 12-18 cm.
ID: rusty-orange thick stalks, pale-greenish polyps

Mushroom xenia coral (*Xenia* sp.)
Xeniidae, 6-9 cm.
ID: milky-white stalk with polyps on the swelling

Sulphur leather coral (*Rhytisma fulvum*)
Leather corals (Alcyoniidae), Indo-Pacific
ID: encrusting colonial species, yellow or grey

SOFT CORALS, SEA FANS, SEA PENS, BLUE CORAL

Mushroom leather coral (*Sarcophyton* sp.)
Leather corals (Alcyoniidae) Indo-Pacific, 50 cm.
On the left: smooth surface with polyps retracted

Finger leather coral (*Sinularia* sp.)
Leather corals (Alcyoniidae) Indo-Pacific, 40 cm.
ID: elongated finger-like branched lobes

Red iciligorgia (*Iciligorgia* cf. *rubra*)
Anthothelidae, 50 cm, West Pacific
ID: red "split" branches, white polyps

White star polyp (*Briareum* sp.)
Briareidae, 40-60 cm (colony)
ID: long brown tentacles, white centres

Branching sea fan (*Melithaea* sp.)
Melithaeidae, West Pacific, 20 cm.
ID: coloration variable, dense branched bushes. IT

Yellow knobby sea fan (Melithaea sp.)
Melithaeidae, West Pacific, 40 cm.
IT

Bicolor sea pen (*Virgularia* sp.)
Veretillidae, Indo-Pacific, 60 cm.
ID: white and purple polyps

Blue coral (*Heliopora coerulea*) Helioporidae
Indo-Pacific. Unique octocoral with massive skeleton,
like hard corals. **ID**: pale polyps with eight tentacles

CNIDARIANS   HEXACORALLIANS

Bulb tentacle anemone (*Entacmaea quadricolor*)
Actiniidae, Red Sea, Indo-Pacific, 30 cm.
ID: long tentacles with bulbs below the tips

Griffiths's fire anemone (*Megalactis griffithsi*)
Actinodendridae
Indo-Pacific, 30 cm

Magnificent anemone (*Heteractis magnifica*)
Stichodactylidae
Red Sea, Indo-Pacific, 30 cm

Leather anemone (*Heteractis crispa*)
Stichodactylidae
Red Sea, Indo-Pacific, 20 cm

Beaded sea anemone (*Phymanthus* cf. *pinnulatum*)
Phymanthidae
Indo-West Pacific, 8 cm

Precious alicia (*Alicia pretiosa*)
Aliciidae, Red Sea, Indo-Pacific, 22 cm.
ID: long transparent stinging tentacles

Swimming anemone (*Boloceroides mcmurrichi*)
Boloceroididae, Indo-Pacific, 8 cm.
ID: numerous striped tentacles (>400)

Club-tentacled anemone (*Telmactis* sp.)
Andvakiidae, West Pacific, 6 cm.
Found on rocky slopes

ANEMONES, WIRE CORALS, CORALLIMORPHS, ZOANTHIDS

Gorgonian wrapper (*Nemanthus annamensis*)
Nemanthidae, Indo-Pacific, 3 cm.
Coloration variable, always in small groups

Black coral whip (*Cirrhipathes* sp.)
Antipathidae, Indo-Pacific, 150 cm. **ID**: pale brown or greenish, with polyps all around the stem

Hoplites corallimorph (*Paracorynactis hoplites*)
Corallimorphidae, IP, 15 cm. Preys on echinoderms, balls (acrospheres) are filled with stinging cells

Elephant ear mushroom coral (*Rhodactis howesii*)
Discosomatidae, Indo-Pacific, 5-8 cm

Flower mushroom coral (*Ricordea yuma*)
Discosomidae, Indo-Pacific, 6 cm
ID: radiating rows of short, berry-shaped tentacles

Graceful hydrozoanthus (*Hydrozoanthus gracilis*)
Hydrozoanthidae, Indo-Pacific, 1 cm (polyp), epibiont on hydrozoan *Plumularia habereri*

Pale zoanthid (*Palythoa* sp.)
Sphenopidae, West Pacific, 3 cm.
ID: long stalks, dense clusters with 12-20 species

Orange zoanthid (*Zoanthus* sp.)
Zoanthidae, West Pacific.
ID: orange oral disc, purple ring around mouth

HEXACORALLIANS ZOANTHIDS, TUBE ANEMONES, HYDROIDS

Encrusting stick anemone (Acrozoanthus sp.)
Zoanthidae
West Pacific, 2 cm

Black-tipped stick anemone (*Acrozoanthus australiae*)
Zoanthidae, Indo-West Pacific, 2,3 cm

Yellow tube anemone (*Cerianthidae* sp.)
Impossible to definitively identify (even to genus) from external features alone

Yellow lace coral (*Distichopora* sp.)
Lace corals (Stylasteridae)
Indo-Pacific, 12 cm

Elegant lace coral (*Stylaster sanguineus*)
Stylasteridae, 8 cm, Indo-Pacific, colours between magenta and white

Orange hydroids (*Eudendrium sp.*)
Eudendriidae, 20 cm, Philippines. Picture shows polyps and white gonophores (reproductive organ)

Stinging hydroid (*Aglaophenia cupressina*)
Aglapheniidae
Indo-Pacific, 35 cm

Branching fire coral (*Millepora sp.*)
Fire corals (Milleporidae)
ID: white-topped, forked at the ends branches

HEXACORALLIANS   HARD CORALS   ACROPORIDAE

Bottlebrush branching coral (*Acropora subglabra*)
Staghorn corals (Acroporidae)
Indo-West Pacific

Purple tipped acropora (*Acropora tenuis*)
Staghorn corals (Acroporidae) Indo-Pacific.
Coloration variable

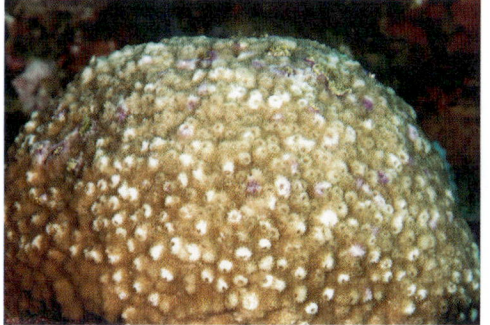

Porous star coral (*Astreopora myriophthalma*)
Staghorn corals (Acroporidae), Indo-West Pacific.
ID: hemispherical, brownish with small purple areas

Column staghorn coral (*Isopora palifera*)
Acroporidae
Indo-Pacific, 100 cm

Obscure montipora (*Montipora confusa*)
Staghorn corals (Acroporidae), Indo-West Pacific.
ID: encrusting base, tower-like branches

Mactan montipora (*Montipora mactanensis*)
Acroporidae
West Pacific

Rainbow montipora (*Montipora* cf. *danae*)
Staghorn corals (Acroporidae)
West Pacific. IT

White-edged montipora (*Montipora* sp.)
Staghorn corals (Acroporidae)
Philippines. IT

193

HARD CORALS — POCILLOPORIIDAE, AGARICIDAE

Verrucose birdnest coral (*Pocillipora verrucosa*)
Pocilloporiidae
Red Sea, Indo-Pacific, 25 cm

Birdsnest coral (*Seriatopora caliendrum*)
Pocilloporiidae
Indo-West Pacific, 35 cm

Hystrix birdsnest coral (*Seriatopora hystrix*)
Pocilloporiidae
Red Sea, Indo-Pacific, 70 cm

Hammer coral (*Euphyllia ancora*)
Euphylliidae
Indo-Pacific, 100 cm (colony)

Gardiner's honeycomb coral (*Gardineroseris planulata*)
Agariciidae, Indo-Pacific, 30-40 cm

Tubular leptoseris (*Leptoseris tubulifera*)
Agariciidae
Indo-Pacific, 30 cm

Fingerprint pavona (*Pavona danai*)
Agariciidae
Indo-West Pacific, 20-30 cm

Veined pavona *(Pavona venosa)*
Agariciidae
Indo-West Pacific, 30 cm

HARD CORALS   FUNGIDAE, MERULINIDAE, PSAMMOCORIDAE

Red disc coral (*Cycloseris* sp.)
Mushroom corals (Fungiidae)
West Pacific, 10 cm

Yellow disc coral (*Cycloseris* sp.)
Mushroom corals (Fungiidae)
West Pacific, 10 cm

Curved trumped coral (*Caulastraea curvata*)
Merulinidae
Indo-Pacific, 15 cm (colony)

Large-polyped brain coral (*Dipsastraea* sp.)
Merulinidae
Philippines, IT, poss. *D.veroni*

Pacific echinopora (*Echinopora pacificus*)
Merulinidae
West Pacific, 30 cm

Big velvet coral (*Hydnophora grandis*)
Merulinidae
Indo-Pacific, 100 cm

Lettuce coral (*Pectinia lactuca*)
Merulinidae
Indo-Pacific, >1 m

Segmented psammogora (*Psammocora haimiana*)
Psammocoridae
Indo-Pacific, 36 cm

HARD CORALS   DENDROPHYLLIIDAE, LOBOPHYLLIIDAE

Cup coral (*Tubastraea* cf. *faulkneri*)
Dendrophylliidae, West Pacific
ID: pink, bright yellow polyps with red centres

Black sun coral (*Tubastraea micranthus*)
Dendrophylliidae
Red Sea, Indo-Pacific, 100+ cm

Greenish disc coral (*Turbinaria frondens*)
Dendrophylliidae, Indo-Pacific.
ID: encrusting brown-greenish cup-shaped colonies

Disc coral (*Turbinaria mesenterina*)
Dendrophylliidae, Indo-Pacific, 60 cm.
ID: corallites only on one side, other side smooth

Bubble coral (*Pleurogyra sinuosa*)
Family uncertain (Scleractinia incertae sedis)
Indo-Pacific, 1 m+ (colony)

Button coral (*Cynarina lacrymalis*)
Lobophylliidae
Indo-Pacific, 20 cm

Red chalice coral *(Echinophyllia* sp., poss. *E.aspera)*
Lobophylliidae
West Pacific, 60 cm

Lobed cactus coral (*Lobophyllia flabelliformis*)
Lobophylliidae
West Pacific. IT

196

HARD CORALS　　　　　　　　　　　　　　　　　　PORITIDAE

Grey-disc alveopora (*Alveopora* sp.)
Poritidae, West Pacific.
ID: grey disc with 12 greenish tentacles

White-disc alveopora (*Alveopora* sp.)
Poritidae, West Pacific.
ID: white disc with 12 brown tentacles

Short-tentacle goniopora (*Goniopora* sp.)
Poritidae, Philippines. IT
ID: white disc with 24 greenish tentacles

Long-tentacle goniopora (*Goniopora* sp.)
Poritidae, Philippines. IT

Branching white-tipped coral (*Porites* sp.)
Poritidae, West Pacific, colonies up to 1 m.
ID: greyish branches, with flat white ends. IT

Rounded white-tipped coral (*Porites* sp.)
Poritidae, West Pacific, colonies up to 1 m.
ID: brownish branches, with rounded white ends. IT

Small knob coral (*Plesiastrea versipora*)
Family uncertain (Scleractinia incertae sedis)
Indo-Pacific, colonies up to 3 m

Red mint blastomussa (*Blastomussa wellsi*)
Family uncertain (Scleractinia incertae sedis)
Indo-Pacific

JELLYFISHES | COMB JELLIES

Moon jellyfish (*Aurelia aurita*)
Scyphozoa (Ulmaridae) Circumglobal, 25 cm.
ID: four horseshoe-shaped gonads (sex glands)

Spotted jellyfish (*Mastigias papua*)
Scyphozoa (Mastigiidae) Indo-Pacific, 15-30 cm.
ID: white spots on the bell and tentacles

Pelagic jellyfish (*Pelagia* sp.)
Scyphozoa (Pelagiidae)
West Pacific, 7 cm. IT

Hat jellyfish (*Olindias* sp.)
Hydrozoa (Olindiidae)
West Pacific, 5 cm. IT

Red-spotted comb jelly (*Coeloplana* sp.,poss *C.* cf. *mellosa*) Coeloplanidae, West Pacific, 3 cm.
Found on soft corals

Soft coral comb jelly (*Coeloplana* sp.)
Coeloplanidae, West Pacific, 1.8 cm.
Found on soft corals

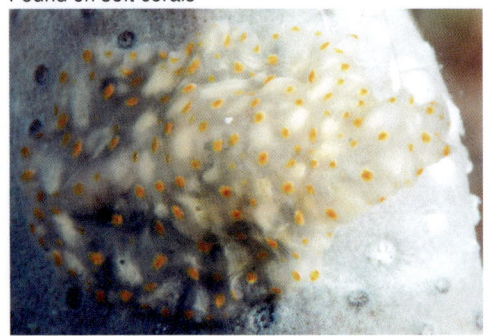

Yellow-spotted comb jelly (*Coeloplana* sp.)
Coeloplanidae, Philippines, 2 cm.
Jellyfish and comb jellies are not relatives

Imperial harp-shaped comb jelly (*Lyrocteis* cf. i*mperatoris*) Lyroctenidae, Indo-Pacific, 20 cm.
Deep dwelling benthic ctenophore. IT

MARINE PLANTS — GREEN ALGAE (CHLOROPHYTA)

Sea grapes (*Caulerpa racemosa*)
Chlorophyta (Caulerpaceae) Indo-Pacific, 1-2 m
spreading colonies, branches up to 30 cm

Cactus tree alga (*Caulerpa serrulata*)
Chlorophyta (Caulerpaceae)
Indo-Pacific, 30 cm

Green feather alga (*Caulerpa sertularioides*)
Chlorophyta (Caulerpaceae)
Circumtropical, 35 cm

Green fingers (*Codium* sp.)
Chlorophyta (Codiaceae)
Philippines, 10 cm

Elephant's ear (*Avrainvillea erecta*) Chlorophyta
(Dichotomosiphonaceae) Indo-Pacific, 12 cm.
Hosts several slugs, incl. *Costasiella kuroshimae*

Watercress alga (*Halimeda opuntia*) Chlorophyta
Halimedaceae
Circumtropical, 30 cm

Green bubble alga (*Dictyosphaeria cavernosa*)
Chlorophyta (Siphonocladaceae)
Circumtropical, 20 cm

Sailor's eyeballs (*Valonia ventricosa*) Chlorophyta
Valoniaceae, circumtropical, 5 cm (single-cell
organism!)

MARINE PLANTS — BROWN ALGAE, RED ALGAE

Gulfweed (*Sargassum* sp.)
Brown algae (Sargassaceae) Indo-West Pacific, 1-2 m.
Bottom right: another brown algae *Padina* sp.

Red sea plume (*Asparagopsis taxiformis*)
Red algae (Bonnemaisoniaceae) Circumtropical, 18 cm. Used as food and medicine

Spreading champia (*Champia expansa*)
Red algae (Champiaceae)
West Pacific, 12 cm

Magnificent hypoglossum (*Hypoglossum* sp.)
Red algae (Delesseriaceae)
Philippines, leafs<1 cm. DBA

Red-top dotyophycus (*Dotyophycus yamadae*)
Red algae (Galaxauraceae)
West Pacific, 12-16 cm

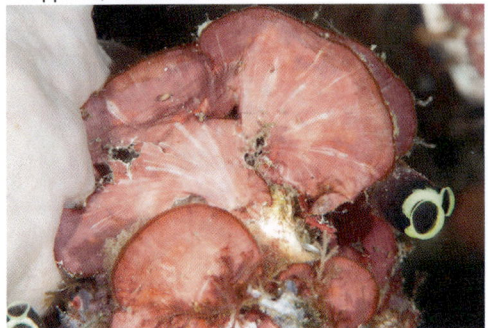

Maroon ethelia (*Ethelia* sp.)
Red algae (Peyssonneliaceae)
Indo-Pacific, 8-12 cm

Chalky red algae (*Titanophora calcarea*)
Red algae (Schizymeniaceae)
West Pacific, 50 cm

Golden-speckled eucheuma (*Eucheuma* sp.)
Red algae (Solieriaceae)
Philippines, 10 cm

REEF ID BOOKS

BOOKSHELF

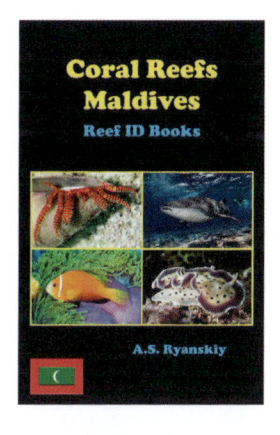

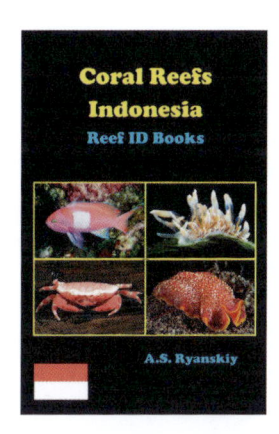

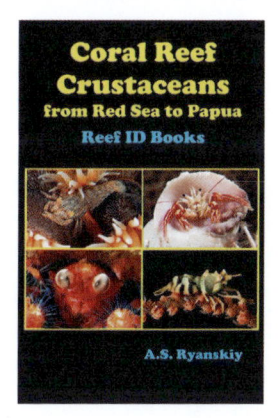

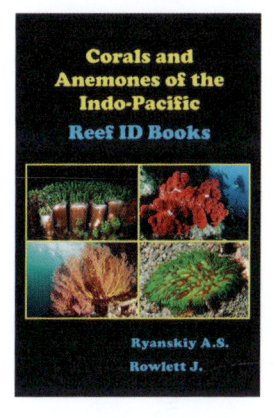

Coral Reefs Maldives — Reef ID Books — A.S. Ryanskiy

Coral Reefs Indonesia — Reef ID Books — A.S. Ryanskiy

Coral Reef Crustaceans from Red Sea to Papua — Reef ID Books — A.S. Ryanskiy

Corals and Anemones of the Indo-Pacific — Reef ID Books — Ryanskiy A.S., Rowlett J.

COMING SOON!

Thank you for taking time to read Coral Reef Philippines. If you enjoyed it, please consider telling your friends and posting a short review on Amazon. Word of mouth is an author's best friend and much appreciated. Thank you! **Andrey Ryanskiy**

Printed in Poland
by Amazon Fulfillment
Poland Sp. z o.o., Wrocław